AESTHETIC MEASURE

LONDON : HUMPHREY MILFORD
OXFORD UNIVERSITY PRESS

AESTHETIC MEASURE

BY

GEORGE D. BIRKHOFF

CAMBRIDGE, MASSACHUSETTS
HARVARD UNIVERSITY PRESS
1933

COPYRIGHT, 1933
BY THE PRESIDENT AND FELLOWS OF HARVARD COLLEGE

PRINTED AT THE HARVARD UNIVERSITY PRESS
CAMBRIDGE, MASSACHUSETTS
UNITED STATES OF AMERICA

To

MARGARET G. BIRKHOFF

WHO ENCOURAGED ME TO FOLLOW FURTHER
THE PROGRAM OF AESTHETIC MEASURE

PREFACE

THE formal structure of Western music began to interest me nearly thirty years ago. It seemed evident that there must exist some guiding principle, and yet no satisfactory explanation of musical form appeared to have been found.

On examining my own limited musical experience, I concluded that the remarkable phenomenon of melody depended mainly upon the orderly arrangement of musical notes in forms appreciated by the ear. Furthermore it looked probable that these relations of order might be classified by empirical methods, and that the best melodies would be found to be characterized by an unusual number of such relations. In this way the theory of aesthetic measure arose in my mind, and the riddle of melody took on the aspect of a quasi-mathematical problem which I, as a mathematician, might profitably study.

Despite the importance and fascination of the investigation so suggested, I put off all further consideration of it in order to give right of way to purely mathematical researches. Only in 1924 did I return to these ideas and try to develop them. It soon appeared that the theory was illustrated most simply by aesthetic objects such as polygons, tilings, and vases, rather than by music. Preliminary results in these fields of geometric form were presented by me in lectures at Pomona, Colorado, and Grinnell Colleges during the same year, and before the Mathematical Association of America at Ithaca in the summer of 1925. In 1928 the Bureau of International Research of Harvard University and Radcliffe College enabled me to spend a half year on leave of absence in the Far East and in Europe, and thus to obtain a more adequate background; to my regret I have not found it feasible to carry out all of the ideas I had then in mind. More recently, with further aid from the Bureau, I have been engaged in making the application of the theory of aesthetic measure to music

PREFACE

itself and to poetry. During all of this period I have presented the results obtained as occasion arose.*

Thus the development of my first ideas concerning music has led me to the daring project of the present book, namely to bring the basic formal side of art within the purview of the simple mathematical formula defining aesthetic measure. This has involved the task of isolating and assessing a large variety of aesthetic factors by the uncertain method of introspection. Any first attempt of this kind is of necessity more or less tentative, and I ask for indulgence on that account.

The psychological basis of the formula, its application to polygonal forms, to ornaments and tilings, and to vases, are developed in the first four chapters.

In Chapters V–VII are taken up the fundamental musical problems related to the diatonic chords, harmony, and melody. In Chapter VIII I have ventured to approach the question of the musical quality in poetry from the same point of view.

A brief review of earlier aesthetic theories is undertaken in Chapter IX, in order to make clear what rôle has been assigned to form in art since the time of Plato. Most of the aesthetic doctrines of the past are found to admit of interpretation in the light of the theory of aesthetic measure. The concluding chapter is devoted to some observations about the general field of art.

Wherever possible I have tried to use the formula in a more or less mechanical fashion for purposes of construction. This has been an interesting and instructive experiment, but the reader will have to judge for himself to what extent the vase forms, simple melodies, and short poem so obtained are successful.

---

* In 1928 I presented at Bologna a first account of the theory of aesthetic measure and its applications to geometric forms; see my article, *Quelques éléments mathématiques de l'art*, Atti del Congresso Internazionale dei Matematici, Bologna, vol. 1, 1929. In 1931 I presented at Paris an outline of the complete project; see my article, *Une théorie quantitative de l'esthétique*, Bulletin de la Société française de Philosophie, 1931. The substance of various chapters has also recently appeared as follows: Chapter I, *A Mathematical Approach to Aesthetics*, Scientia, 1931; Chapter II, *Polygonal Forms*, Sixth Yearbook of the National Council of Teachers of Mathematics, New York, 1931; Chapters V–IX, *A Mathematical Theory of Aesthetics and its Applications to Poetry and Music*, The Rice Institute Pamphlet, vol. 19, 1932.

PREFACE

Of course it would be absurd to think of the basic mathematical formula as a *deus ex machina*, and thus affirm with the Doctor in 'Frier Bacon and Frier Bongay':

> For he that reades but Mathematicke rules
> Shall finde conclusions that availe to worke
> Wonders that passe the common sense of men.

The true function of the concept of aesthetic measure is to provide systematic means of analysis in simple formal aesthetic domains. There is a vast difference between the discovery of a diamond and its appraisal; still more, between the creation of a work of art and an analysis of the formal factors which enter into it.

GEORGE D. BIRKHOFF

CAMBRIDGE, MASSACHUSETTS
October, 1932

CONTENTS

CHAPTER I

THE BASIC FORMULA 3

 1. The Aesthetic Problem. — 2. Nature of the Aesthetic Experience. — 3. Mathematical Formulation of the Problem. — 4. The Feeling of Effort in Aesthetic Experience. — 5. The Psychological Meaning of 'Complexity.' — 6. Associations and Aesthetic Feeling. — 7. The Intuitive Nature of Such Associations. — 8. The Rôle of Sensuous Feeling. — 9. Formal and Connotative Associations. — 10. Formal and Connotative Elements of Order. — 11. Types of Formal Elements of Order. — 12. The Psychological Meaning of 'Order.' — 13. The Concept of Aesthetic Measure. — 14. The Basic Formula. — 15. A Mathematical Argument. — 16. The Scope of the Formula. — 17. A Diagram. — 18. The Method of Application.

CHAPTER II

POLYGONAL FORMS 16

 1. Polygons as Aesthetic Objects. — 2. Preliminary Requirements. — 3. Symmetry of Polygons. — 4. Isosceles and Equilateral Triangles. — 5. Scalene Triangles. — 6. Conclusions concerning Triangular Forms. — 7. Plato's Favorite Triangle. — 8. The Scalene Triangle in Japanese Art. — 9. Symmetric Quadrilaterals (First Type). — 10. Symmetric Quadrilaterals (Second Type). — 11. Symmetric Quadrilaterals (Third Type). — 12. Asymmetric Quadrilaterals. — 13. Conclusions concerning Quadrilateral Forms. — 14. The 'Golden Rectangle' and Others. — 15. Comparison of Rectangular Forms. — 16. Use of Rectangular Forms in Composition. — 17. The Forms of Five- and Six-sided Polygons. — 18. More Complicated Forms. — 19. On the Structure of the Aesthetic Formula. — 20. The Complexity C. — 21. The Element V of Vertical Symmetry. — 22. The Element E of Equilibrium. — 23. The Group of Motions of a Polygon. — 24. The Element R of Rotational Symmetry. — 25. The Element HV of Relation to a Horizontal-Vertical Network. — 26. The Element F of Unsatisfactory Form. — 27. Recapitulation of Definition of Aesthetic Measure. — 28. Application to 90 Polygons. — 29. Possible Modifications. — 30. The Mathematical Treatment of Aesthetic Questions. — 31. On Uncertainty and Optical Illusions.

CONTENTS

CHAPTER III

ORNAMENTS AND TILINGS 49

1. Ornaments and Motions. — 2. Quasi-Ornaments. — 3. The Group of Motions of an Ornament. — 4. Classification of Ornaments. — 5. On the Determination of Species. — 6. Ornaments and Ornamental Patterns. — 7. The Aesthetic Problem of Ornaments. — 8. Polygons as Simple Rectilinear Ornaments. — 9. Aesthetic Measure of Simple Rectilinear Ornaments. — 10. Some Explanatory Comments. — 11. Application to 30 Simple Rectilinear Ornaments. — 12. Other Types of Ornaments. — 13. The Aesthetic Measure of Tilings. — 14. Application to 12 Tilings.

CHAPTER IV

VASES 67

1. The Problem of the Vase. — 2. The Method of Attack. — 3. The Symbol of the Vase. — 4. Characteristic Points and Tangents. — 5. The Appreciable Elements of Order. — 6. The Problem of Regularity of Contour. — 7. Curvature of Contour. — 8. Requirements for Regularity of Contour. — 9. Conventional and Utilitarian Requirements. — 10. On the Interpretation of Vase Form. — 11. The Complexity C of Vase Form. — 12. The Order O of Vase Form. — 13. The Aesthetic Measure M of Vase Form. — 14. Application to Some Chinese Vases. — 15. A Detailed Analysis. — 16. Three Experimental Vases. — 17. General Conclusions.

CHAPTER V

THE DIATONIC CHORDS 87

1. The Problem of Musical Form. — 2. Harmony as an Aesthetic Factor. 3. Consonance and Dissonance. — 4. Musical Notes, Intervals, Triads, and Chords. — 5. The Natural Diatonic Scale. — 6. The Equally Tempered Diatonic Scale. — 7. The Chromatic Scale. Tonality. Modulation. — 8. Major and Minor Modes. — 9. The Problem of Harmony. The Single Chord. — 10. The Intervals. — 11. The Major, Minor, and Diminished Triads. — 12. The Corresponding Chords. — 13. The Dominant 7th and 9th Chords. — 14. The Higher Dominant Chords. — 15. The Derivatives of Dominant Chords. — 16. Regular and Irregular Chords. — 17. Final Classification of Regular Chords. — 18. Incomplete and Ambiguous Regular Chords. — 19. The Chord Value Cd. — 20. The Interval Value I. — 21. The Value D of the Dominating Note. — 22. The Aesthetic Measure m of the Single Chord. — 23. The Complete Triads in Root Position. — 24. The Incomplete Triads in Root Position. — 25. The First Inversions of the Triads. — 26. The Second Inversions of the Triads. — 27. The Dominant

CONTENTS

7th Chord. — 28. The Dominant 9th Chord. — 29. The Dominant 11th Chord. — 30. The Dominant 13th Chord. — 31. The Derivative Chords. — 32. Irregular Chords. — 33. Summary.

CHAPTER VI

DIATONIC HARMONY 128

1. The Problem of Chordal Sequences. — 2. The Method of Attack. — 3. Further Limitation of the Problem. — 4. The Limitation of Leaps. — 5. Limitation of Similarity of Function. — 6. The Law of Resolution. — 7. The Element of Resolution: $R = 4, -4$. — 8. The Cadential Element: $Cl = 4, -2$. — 9. The Element of Dominant Sequence: $D = 4$. — 10. The Elements $SF = 4$ and $RF = 4$. — 11. The Element of Progression: $P = 2, 0, -2$. — 12. The Negative Element $FR = 4$. — 13. The Negative Element of the Mediant: $Mt = 2, 4$. — 14. The Negative Element of the Leading Note: $LN = 2, 4$. — 15. The Negative Element of Stationary Notes: $SN = 2, 4$. — 16. The Negative Element of Dissonant Leap: $DL = 2$. — 17. The Negative Element of Similar Motion: $SM = 2$. — 18. The Negative Element of Bass Leap: $BL = 4$. — 19. Recapitulation of Definition of M for Chordal Sequences. — 20. Comparison with Prout's Classification. — 21. General Remarks.

CHAPTER VII

MELODY 152

1. Introduction. — 2. The Problem of Melody. — 3. The Rôle of Harmony. — 4. The Question of Phrasing and Comparison. — 5. Limitation of Leaps. — 6. Preparatory Analysis of a Beethoven Chorale. — 7. Definition of O, C, and M for Simple Melody. — 8. Further Conditions of Satisfactory Form. — 9. The Application to Melody. — 10. First Experimental Melody. — 11. Second Experimental Melody. — 12. Third Experimental Melody. — 13. The Problem of Rhythm.

CHAPTER VIII

THE MUSICAL QUALITY IN POETRY 170

1. The Tripartite Nature of Verse. — 2. The Musical Quality in Poetry. — 3. Rhyme. — 4. Assonance. — 5. Alliteration. — 6. The Musical Vowel Sounds. — 7. 'Anastomosis.' — 8. Poe's Concept of Verse. — 9. Sylvester's Concept of Verse. — 10. On Phonetic Analysis. — 11. The Complexity C. — 12. The Element $2r$ of Rhyme. — 13. The Element aa of Alliteration and Assonance. — 14. The Element $2ae$ of Alliterative and Assonantal Excess. — 15. The Element $2m$ of Musical Vowels. — 16. The Element $2ce$

CONTENTS

of Consonantal Excess. — 17. The Aesthetic Formula. — 18. Analysis of Five Verses of 'Kubla Khan.' — 19. An Experimental Poem. — 20. Further Examples. — 21. Sonorous Prose. — 22. Poetry in Other Languages. — 23. The Rôle of Musical Quality in Poetry.

CHAPTER IX

Earlier Aesthetic Theories 191

1. Introduction. — 2. Plato. — 3. Aristotle. — 4. Plotinus. — 5. The Greek View. — 6. Luca Paciolo. Michelangelo. — 7. Fracastoro. — 8. Wit and Taste. — 9. Descartes. — 10. Leibnitz. — 11. Boileau. De Crousaz. — 12. Vico. — 13. Rameau. — 14. Euler. — 15. Hogarth. — 16. Burke. — 17. Hemsterhuis. — 18. Kant. — 19. Schiller. Hegel. — 20. Herbart. — 21. Schleiermacher. — 22. Poe. — 23. Spencer. — 24. Helmholtz. — 25. Sylvester. — 26. Hanslick. — 27. Fechner. — 28. Lanier. — 29. Lipps. — 30. Gurney. — 31. Croce. — 32. Ross. Pope. — 33. The Eastern View of Art. — 34. Concluding Remarks.

CHAPTER X

Art and Aesthetics 209

1. Types of Aesthetic Experience. Art. — 2. The Variability of Aesthetic Values. — 3. Qualitative and Quantitative Aesthetics. — 4. The Qualitative Application of the Aesthetic Formula. — 5. Decorative Design. — 6. Painting, Sculpture, and Architecture. — 7. Western Music. — 8. Eastern Music. — 9. Evolution in Art. — 10. Creative Art.

Index 219

LIST OF PLATES

I. Floor Mosaic Detail from Santa Maria Maggiore, Rome . . 16
II–VII. Aesthetic Measures of 90 Polygonal Forms (in color) . . 32
VIII. The 7 Species of One-Dimensional Ornaments 56
IX–XII. The 17 Species of Two-Dimensional Ornaments 56
XIII–XIV. Aesthetic Measures of 30 Simple Ornaments 60
XV. Aesthetic Measures of 12 Tilings 64
XVI–XVII. Aesthetic Measures of 8 Chinese Vases 80
XVIII. Illustrative Vase Form (in color) 82
XIX. First Experimental Vase Form (in color) 84
XX. Second Experimental Vase Form (in color) 84
XXI. Third Experimental Vase Form (in color) 84
XXII. Correggio's 'Danae' (with linear analysis); Veronese's 'Family of Darius before Alexander' 212
XXIII. South Façade of Taj Mahal, Agra 214

AESTHETIC MEASURE

CHAPTER I

THE BASIC FORMULA

1. The Aesthetic Problem

MANY auditory and visual perceptions are accompanied by a certain intuitive feeling of value, which is clearly separable from sensuous, emotional, moral, or intellectual feeling. The branch of knowledge called aesthetics is concerned primarily with this aesthetic feeling and the aesthetic objects which produce it.

There are numerous kinds of aesthetic objects, and each gives rise to aesthetic feeling which is *sui generis*. Such objects fall, however, in two categories: some, like sunsets, are found in nature, while others are created by the artist. The first category is more or less accidental in quality, while the second category comes into existence as the free expression of aesthetic ideals. It is for this reason that art rather than nature provides the principal material of aesthetics.

Of primary significance for aesthetics is the fact that the objects belonging to a definite class admit of direct intuitive comparison with respect to aesthetic value. The artist and the connoisseur excel in their power to make discriminations of this kind.

To the extent that aesthetics is successful in its scientific aims, it must provide some rational basis for such intuitive comparisons. In fact it is the fundamental problem of aesthetics to determine, within each class of aesthetic objects, those specific attributes upon which the aesthetic value depends.

2. Nature of the Aesthetic Experience

The typical aesthetic experience may be regarded as compounded of three successive phases: (1) a preliminary effort of attention, which is necessary for the act of perception, and which increases in proportion to what we shall call the *complexity* (C) of the object; (2) the feeling of value or *aesthetic measure* (M) which rewards this effort; and finally (3) a realization that the object is characterized by a certain harmony, symmetry, or *order* (*O*), more or less concealed, which seems necessary to the aesthetic effect.

3. Mathematical Formulation of the Problem

This analysis of the aesthetic experience suggests that the aesthetic feelings arise primarily because of an unusual degree of harmonious interrelation within the object. More definitely, if we regard *M*, *O*, and *C* as measurable variables, we are led to write

$$M = \frac{O}{C}$$

and thus to embody in a basic formula the conjecture that the aesthetic measure is determined by the density of order relations in the aesthetic object.

The well known aesthetic demand for 'unity in variety' is evidently closely connected with this formula. The definition of the beautiful as that which gives us the greatest number of ideas in the shortest space of time (formulated by Hemsterhuis in the eighteenth century) is of an analogous nature.

If we admit the validity of such a formula, the following mathematical formulation of the fundamental aesthetic problem may be made: *Within each class of aesthetic objects, to define the order* O *and the complexity* C *so that their ratio* $M = O/C$ *yields the aesthetic measure of any object of the class.*

It will be our chief aim to consider various simple classes of aesthetic objects, and in these cases to solve as best we can the fundamental aesthetic problem in the mathematical form just stated. Preliminary to such actual application, however, it is desirable to indicate the psychological basis of the formula and the conditions under which it can be applied.

4. The Feeling of Effort in Aesthetic Experience

From the physiological-psychological point of view, the act of perception of an aesthetic object begins with the stimulation of the auditory or visual organs of sense, and continues until this stimulation and the resultant cerebral excitation terminate. In order that the act of perception be successfully performed, there is also required the appropriate field of attention in consciousness. The attentive attitude has of course its physiological correlative, which in particular ensures that the motor adjustments requisite to the act of perception are effected when required. These adjustments are usually made without the intervention of motor ideas such as accompany all voluntary motor acts, and in this sense are 'automatic.' In more physiological terms, the stimulation sets up a nerve current which, after reaching the cerebral cortex, in part reverts to the periphery as a motor nerve current along a path of extreme habituation, such as corresponds to any automatic act.

Now, although these automatic adjustments are made without the intervention of motor ideas, nevertheless there is a well known feeling of effort or varying tension while the successive adjustments are called for and performed. This constitutes a definite and important part of the general feeling characteristic of the state of attention. The fact that interest of some kind is almost necessary for sustained attention would seem to indicate that this feeling has not a positive (pleasurable) tone but rather a negative one. Furthermore, if we bear in mind that the so-called automatic acts are nothing but the outcome of unvarying voluntary acts habitually performed, we may reasonably believe that there remain vestiges of the motor ideas originally involved, and that it is these which make up this feeling of effort.

From such a point of view, the feeling of effort always attendant upon perception appears as a summation of the feelings of tension which accompany the various automatic adjustments.

5. The Psychological Meaning of 'Complexity'

Suppose that $A, B, C, \ldots$ are the various automatic adjustments required, with respective indices of tension $a, b, c, \ldots$, and that these adjustments $A, B, C, \ldots$ take place $r, s, t, \ldots$ times respectively. Now it is the feeling of effort or tension which is the psychological counterpart of what has been referred to as the complexity C of the aesthetic object. In this manner we are led to regard the sum of the various indices as the measure of complexity, and thus to write

$$C = ra + sb + tc + \cdots.$$

A simple illustration may serve to clarify the point of view. Suppose that we fix attention upon a convex polygonal tile. The act of perception involved is so quickly performed as to seem nearly instantaneous. The feeling of effort is almost negligible while the eye follows the successive sides of the polygon and the corresponding motor adjustments are effected automatically. Nevertheless, according to the point of view advanced above, there is a slight feeling of tension attendant upon each adjustment, and the complexity C will be measured by the number of sides of the polygon.

Perhaps a more satisfying illustration is furnished by any simple melody. Here the automatic motor adjustments necessary to the act of perception are the incipient adjustments of the vocal cords to the successive tones. Evidently in this case the complexity C will be measured by the number of notes in the melody.

6. Associations and Aesthetic Feeling

Up to this point we have only considered the act of perception of an aesthetic object as involving a certain effort of attention. This feeling of effort is correlated with the efferent part of the nerve current which gives rise to the required automatic motor adjustments, and has no direct reference to aesthetic feeling.

For the cause (physiologically speaking) of aesthetic feeling, we must look to that complementary part of the nerve current which, impinging on the auditory and visual centers, gives rise to sensations derived from the object, and, spreading from thence, calls various associated ideas with their attendant feelings into play. These sensations, together with the associated ideas and their attendant feelings, constitute the full perception of the object. It is in these associations rather than in the sensations themselves that we shall find the determining aesthetic factor.

In many cases of aesthetic perception there is more or less complete identification of the percipient with the aesthetic object. This feeling of 'empathy,' whose importance has been stressed by the psychologist Lipps,* contributes to the enhancement of the aesthetic effect. Similarly,

* *Ästhetik: Psychologie des Schönen und der Kunst*, Hamburg and Leipzig, vol. 1 (1903), vol. 2 (1906).

actual participation on the part of the percipient, as in the case of singing a tune as well as hearing it, will enhance the effect.

7. The Intuitive Nature of Such Associations

Mere verbal associations are irrelevant to the aesthetic experience. In other words, aesthetic associations are *intuitive* in type.

When, for instance, I see a symmetrical object, I feel its pleasurable quality, but do not need to assert explicitly to myself, "How symmetrical!" This characteristic feature may be explained as follows. In the course of individual experience it is found generally that symmetrical objects possess exceptional and desirable qualities. Thus our own bodies are not regarded as perfectly formed unless they are symmetrical. Furthermore, the visual and tactual technique by which we perceive the symmetry of various objects is uniform, highly developed, and almost instantaneously applied. It is this technique which forms the associative 'pointer.' In consequence of it, the perception of any symmetrical object is accompanied by an intuitive aesthetic feeling of positive tone.

It would even seem to be almost preferable that no verbal association be made. The unusual effectiveness of more or less occult associations in aesthetic experience is probably due to the fact that such associations are never given verbal reference.

8. The Rôle of Sensuous Feeling

The typical aesthetic perception is primarily of auditory or visual type, and so is not accompanied by stimulation of the end-organs of the so-called lower senses. Thus the sensuous feeling which enters will be highly refined. Nevertheless, since sensuous feeling with a slight positive tone ordinarily accompanies sensations of sight and of sound, it might appear that such sensuous feeling requires some consideration as part of the aesthetic feeling. Now, in my opinion, this component can be set aside in the cases of most interest just because the positive tone of sensuous feeling is always present, and in no way differentiates one perception from another.

For example, all sequences of pure musical tones are equally agreeable as far as the individual sensations are concerned. Yet some of these sequences are melodic in quality, while others are not. Hence, although the agreeableness of the individual sounds forms part of the tone of feeling, we may set aside this sensuous component when we compare the melodic quality of various sequences of musical tones.

To support this opinion further, I will take up briefly certain auditory facts which at first sight appear to be in contradiction with it.

If a dissonant musical interval, such as a semitone, is heard, the resultant tone of feeling is negative. Similarly, if a consonant interval like the perfect fifth is heard, the resultant tone of feeling is positive. But is not the sensation of a dissonant interval to be considered a single auditory sensation comparable with that of a consonant interval, and is it not necessary in this case at least to modify the conclusion as to the constancy of the sensuous factor?

In order to answer this question, let us recall that musical tones, as produced either mechanically or by the human voice, contain a pure fundamental tone of a certain frequency of vibration and pure overtones of double the frequency (the octave), of triple the frequency (the octave of the perfect fifth), etc.; here, with Helmholtz, we regard a pure tone as the true individual sensation of sound. Thus 'association by contiguity' operates to connect any tone with its overtones.

If such be the case, a dissonant interval, being made up of two dissociated tones, may possess a negative tone of feeling on account of this dissociation; while the two constituent tones of a consonant interval, being connected by association through their overtones, may possess a positive tone of feeling in consequence. Hence the obvious difference in the aesthetic effect of a consonant and a dissonant musical interval can be explained on the basis of association alone.

9. Formal and Connotative Associations

It is necessary to call attention to a fundamental division of the types of associations which enter into the aesthetic experience.

Certain kinds of associations are so simple and unitary that they can be at once defined and their rôle can be ascertained with accuracy. On the other hand, there are many associations, of utmost importance from the aesthetic point of view, which defy analysis because they touch our experience at so many points. The associations of the first type are those such as symmetry; an instance of the second type would be the associations which are stirred by the *meaning* of a beautiful poem.

For the purpose of convenient differentiation, associations will be called 'formal' or 'connotative' according as they are of the first or second type. There will of course be intermediate possibilities.

More precisely, formal associations are such as involve reference to some simple physical property of the aesthetic object. Two simple instances of these are the following:

>rectangle in vertical position → symmetry about vertical;
>interval of note and its octave → consonance.

There is no naming of the corresponding property, which is merely pointed out, as it were, by the visual or auditory technique involved.

All associations which are not of this simple formal type will be called connotative.

10. Formal and Connotative Elements of Order

The property of the aesthetic object which corresponds to any association will be called an 'element of order' in the object; and such an element of order will be called formal or connotative according to the nature of the association. Thus a formal element of order arises from a simple physical property such, for instance, as that of consonance in the case of a musical interval or of symmetry in the case of a geometrical figure.

It is not always the case that the elements of order and the corresponding associations are accompanied by a positive tone of feeling. For example, sharp dissonance is to be looked upon as an element of order with a negative tone of feeling.

11. Types of Formal Elements of Order

The actual types of formal elements of order which will be met with are mainly such obvious positive ones as repetition, similarity, contrast, equality, symmetry, balance, and sequence, each of which takes many forms. These are in general to be reckoned as positive in their effect.

Furthermore there is a somewhat less obvious positive element of order, due to suitable centers of interest or repose, which plays a rôle. For example, a painting should have one predominant center of interest on which the eye can rest; similarly in Western music it is desirable to commence in the central tonic chord and to return to this center at the end.

On the other hand, ambiguity, undue repetition, and unnecessary imperfection are formal elements of order which are of strongly negative type. A rectangle nearly but not quite a square is unpleasantly ambiguous; a poem overburdened with alliteration and assonance fatigues by undue repetition; a musical performance in which a single wrong note is heard is marred by the unnecessary imperfection.

12. The Psychological Meaning of 'Order'

We are now prepared to deal with the order O of the aesthetic object in a manner analogous to that used in dealing with the complexity C.

Let us suppose that associations of various types $L, M, N, \ldots$ take place with respective indices of tone of feeling $l, m, n, \ldots$. In this case the indices may be positive, zero, or negative, according as the corresponding tones of feeling are positive, indifferent, or negative. If the associations $L, M, N, \ldots$ occur $u, v, w, \ldots$ times respectively, then we may regard the total tone of feeling as a summational effect represented by the sum $ul + vm + \cdots$.

This effect is the psychological counterpart of what we have called the order O of the aesthetic object, inasmuch as $L, M, N, \ldots$ correspond to what have been termed the elements of order in the aesthetic object. Thus we are led to write

$$O = ul + vm + wn + \cdots.$$

By way of illustration, let us suppose that we have before us various polygonal tiles in vertical position. What are the elements of order and the corresponding associations which determine the feeling of aesthetic value accompanying the act of perception of such a tile? Inasmuch as a detailed study of polygonal form is made in the next chapter, we shall merely mention three obvious positive elements of order, without making any attempt to choose indices. If a tile is symmetric about a vertical axis, the vertical symmetry is felt pleasantly. Again, a tile may have symmetry of rotation; a square tile, for example, has this property, for it can be rotated through a right angle without affecting its position. Such symmetry of rotation is also appreciated immediately. Lastly, if the sides of a tile fall along a rectangular network, as in the case of a Greek cross, the relation to the network is felt agreeably.

13. The Concept of Aesthetic Measure

The aesthetic measure M of a class of aesthetic objects is primarily any quantitative index of their comparative aesthetic effectiveness.

It is impossible to compare objects of different types, as we observed at the outset. Who, for instance, would attempt to compare a vase with a melody? In fact, for comparison to be possible, such classes must be severely restricted. Thus it is futile to compare a painting in oils with one in water colors, except indirectly, by the comparison of each with the best examples of its type; to be sure, the two paintings might be compared, in respect to composition alone, by means of photographic reproduction. On the other hand, photographic portraits of the same person are readily compared and arranged in order of preference.

But even when the class is sufficiently restricted, the preferences of different individuals will vary according to their taste and aesthetic experience. Moreover the preference of an individual will change somewhat from time to time. Thus such aesthetic comparison, of which the aesthetic measure M is the determining index, will have substantial meaning only when it represents the normal or average judgment of some selected group of observers. For example, in the consideration of Western music it would be natural to abide by the consensus of opinion of those who are familiar with it.

Consequently the concept of aesthetic measure M is applicable only if the class of objects is so restricted that direct intuitive comparison of the different objects becomes possible, in which case the arrangement in order of aesthetic measure represents the aesthetic judgment of an idealized 'normal observer.'

14. The Basic Formula

If our earlier analysis be correct, it is the intuitive estimate of the amount of order O inherent in the aesthetic object, as compared with its complexity C, from which arises the derivative feeling of the aesthetic measure M of the different objects of the class considered. We shall first make an argument to this effect on the basis of an analogy, and then proceed to a more purely mathematical argument.

The analogy will be drawn from the economic field. Among business enterprises of a single definite type, which shall be held the most successful? The usual answer would take the following form. In each business there is involved a certain investment i and a certain annual profit p. The ratio p/i, which represents the percentage of interest on the investment, is regarded as the economic measure of success.

Similarly in the perception of aesthetic objects belonging to a definite class, there is involved a feeling of effort of attention, measured by C, which is rewarded by a certain positive tone of feeling, measured by O. It is natural that reward should be proportional to effort, as in the case of a business enterprise. By analogy, then, it is the ratio O/C which best represents the aesthetic measure M.

15. A Mathematical Argument *

More mathematically, but perhaps not more convincingly, we can argue as follows. In the first place it must be supposed that if two objects of the class have the same order O and the same complexity C, their aesthetic measures are to be regarded as the same. Hence we may write

$$M = f(O, C)$$

and thus assert that the aesthetic measure depends functionally upon O and C alone.

It is obvious that if we increase the order without altering the complexity, or if we diminish the complexity without altering the order, the value of M should be increased. But these two laws do not serve to determine the function f.

In order to do so, we imagine the following hypothetical experiment. Suppose that we have before us a certain set of k objects of the class, all having the same order O and the same complexity C, and also a second set of k' objects of the class, all having the order O' and complexity C'. Let us choose k and k' so that $k'C'$ equals kC.

* This mathematical section may be omitted.

THE BASIC FORMULA

Now proceed as follows. Let all of the first set of objects be observed, one after the other; the total effort will be measured by kC of course, and the total tone of aesthetic feeling by kO. Similarly let all of the second set be observed. The effort will be the same as before, since $k'C'$ equals kC; and the total tone of feeling will be measured by $k'O'$.

If the aesthetic measure of the individual objects of the second class is the same as of the first, it would appear inevitable that the total tone of feeling must be the same in both cases, so that $k'O'$ equals kO. With this granted, we conclude at once that the ratios O'/C' and O/C are the same. In consequence the aesthetic measure only depends upon the ratio O to C:

$$M = f\left(\frac{O}{C}\right).$$

The final step can now be taken. Since it is not the actual numerical magnitude of f that is important but only the relative magnitude when we order according to aesthetic measure, and since M must increase with O/C, we can properly define M as equal to the ratio of O to C.

It is obvious that the aesthetic measure M as thus determined is zero ($M = 0$) when the tone of feeling due to the associated ideas is indifferent.

16. THE SCOPE OF THE FORMULA

As presented above, the basic formula admits of theoretic application to any properly restricted class of aesthetic objects.

Now it would seem not to be difficult in any case to devise a reasonable and simple measure of the complexity C of the aesthetic objects of the class. On the other hand, the order O must take account of all types of associations induced by the objects, whether formal or connotative; and a suitable index is to be assigned to each. Unfortunately the connotative elements of order cannot be so treated, since they are of inconceivable variety and lie beyond the range of precise analysis.

It is clear then that complete quantitative application of the basic formula can only be effected when the elements of order are mainly formal. Of course it is always possible to consider the formula only in so far as the formal elements of order are concerned, and to arrive in this way at a partial application.

AESTHETIC MEASURE

Consequently our attention will be directed almost exclusively towards the formal side of art, to which alone the basic formula of aesthetic measure can be quantitatively applied. Our first and principal aim will be to effect an analysis in typical important cases of the utmost simplicity. From the vantage point so reached it will be possible to consider briefly more general questions. In following this program, there is of course no intention of denying the transcendent importance of the connotative side in all creative art.

17. A Diagram

The adjoining diagram with the attached legend may be of assistance in recalling the above analysis of the aesthetic experience and the basic aesthetic formula to which it leads.

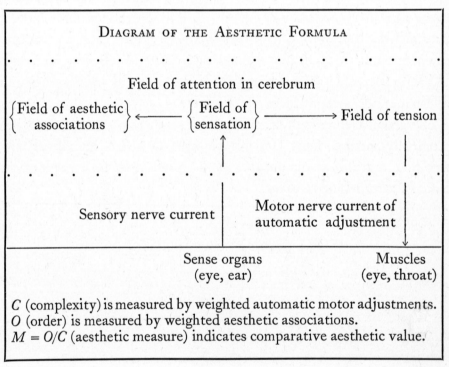

Figure 1

18. The Method of Application

Even in the most favorable cases, the precise rules adopted for the determination of O, C, and thence of the aesthetic measure M, are necessarily empirical. In fact the symbols O and C represent social values, and share in the uncertainty common to such values. For example, the 'purchasing power of money' can only be determined approximately by means of empirical rules, and yet the concept involved is of fundamental economic importance.

At the same time it should be added that this empirical method seems to be the only one by which concepts of this general category can be approached scientifically.

We shall endeavor at all times to choose formal elements of order having unquestionable aesthetic importance, and to define indices in the most simple and reasonable manner possible. The underlying facts have to be ascertained by the method of direct introspection.

In particular we shall pay attention to the two following *desiderata*:

As far as possible these indices are to be taken as equal, or else in the simplest manner compatible with the facts.

The various elements of order are to be considered only in so far as they are logically independent. If, for example, $a = b$ is an equality which enters in O, and if $b = c$ is another such equality, then the equality $a = c$ will not be counted separately.

CHAPTER II

POLYGONAL FORMS

1. Polygons as Aesthetic Objects

POLYGONS, considered merely in their aspect of geometric form, have a definite, if elementary, aesthetic appeal. This fact has always been recognized, and is borne out by their wide use for decorative purposes in East and West (see Plate I opposite). Moreover such polygonal forms can be intuitively compared with one another with respect to aesthetic quality. For example, Alison says:* "An Equilateral Triangle is more beautiful than a Scalene or an Isosceles, a Square than a Rhombus, an Hexagon than a Square, an Ellipse than a Parabola, a Circle than an Ellipse; because the number of their uniform parts are greater, and their Expression of Design more complete."

Evidently then polygonal forms constitute a class of aesthetic objects of the utmost simplicity, which have the further advantage of being relatively free from connotative elements of order. It is for these reasons that the first application of the general theory is made to polygonal forms.

2. Preliminary Requirements

According to the general theory, it is necessary to select some specific type of representation of the polygons. For the sake of definiteness we shall have in mind porcelain tiles of polygonal shape, and alike in size, color, and material. In this way the class of objects to be considered is precisely defined. Other classes of polygonal objects might be considered, such for instance as precious stones cut in polygonal form. But it is evident that then various factors other than form would be likely to enter, such as the brilliancy of the reflected light. Thus the choice of tiles is advantageous, since these differ from one another only in their aspect of geometric form.

* *Essays on the Nature and Principles of Taste*, Edinburgh (1790).

Floor Mosaic Detail from Santa Maria Maggiore, Rome

POLYGONAL FORMS

A further requirement must be imposed in order to fix the psychological condition of the 'normal observer.' Such a polygonal tile produces a somewhat different impression when it is seen upon a table than when it is seen in vertical position. In fact a tile lying upon the table would be viewed from various directions, while one in vertical position is seen in a single orientation. Therefore it is desirable to think of such tiles as in vertical position. In general the selected orientation will of course be the best one. Perhaps the actual use for decorative purposes which most nearly conforms to these conditions is that in which identical porcelain tiles appear at regular intervals in the same orientation along a stuccoed wall.

Just as in other aesthetic fields, a certain degree of familiarity with the various types of objects involved is required before the aesthetic judgment becomes certain. It is hardly necessary to observe that when novel polygons, pleasing in themselves, are seen for the first time, they take higher rank than they do subsequently, just because of this novelty.

It is clear that when these requirements are satisfied, the aesthetic problem of polygonal form becomes a legitimate one in the sense of the preceding chapter.

3. Symmetry of Polygons

It is desirable at the outset to obtain a clear idea of the types of symmetry which occur in polygonal forms, since such symmetry is evidently of fundamental importance from the aesthetic point of view.

There are two types of symmetry which a polygon may possess. The first and simpler is that of 'symmetry about a line' in its plane, called the 'axis of symmetry.' If the figure be rotated about this line through an angle of 180°, it returns as a whole to its initial position, while the individual points are transferred to the symmetric points on the other side of the axis of symmetry.

In a square the two diagonals are axes of symmetry as well as the two lines through its center which are parallel to a pair of sides. In a rectangle there are only two axes of symmetry, namely the two lines through its center parallel to a pair of sides. A parallelogram is not symmetric about any line through its center.

The second type is that of 'rotational symmetry about a point' in the plane of the polygon, called the 'center of symmetry.' If the polygon be rotated in the plane through a certain angle about this point, it returns as a whole to its initial position, while the individual points are rotated through this angle about the point. The angle in question will be called the 'angle of rotation.' Here the axis of rotation is a line perpendicular to the plane of the polygon at the center of symmetry.

The square possesses such rotational symmetry about its center, and the angle of rotation is clearly 90° or one quarter of a complete revolution. Similarly the rectangle and parallelogram have rotational symmetry with 180° as angle of rotation.

An isosceles triangle illustrates the fact that a polygon may possess symmetry about an axis without having rotational symmetry; and the case of the parallelogram shows that it may possess rotational symmetry without having symmetry about an axis.

By the angle of rotation we mean of course the least such angle. Suppose that q successive rotations through this least angle effect one complete revolution. The least angle of rotation is then $360°/q$. But twice this angle, three times this angle, and, more generally, any multiple of it, are admissible angles also.

When q is an even number, $q/2$ rotations through the least angle will amount to a half revolution or 180°. A polygon with an admissible angle of rotation of 180° is said to possess 'central symmetry.'

In this case any point of the polygon is paired with an opposite corresponding point, so that the line joining the two points is bisected by the center of symmetry. Such central symmetry exists if, and only if, an even number of rotations, q, through the angle of rotation is required before a complete revolution is effected. Evidently the rectangle and parallelogram with q equal to 2, and the square with q equal to 4, have central symmetry, while the equilateral triangle with q equal to 3 has not, in accordance with the statement just made.

It is clear that the above definitions apply to all plane geometrical figures as well as to polygons. In the exceptional case of the circle, every line through the center is an axis of symmetry, and every angle is an admissible angle of rotation. This is not true of any plane figure other than the circle.

4. Isosceles and Equilateral Triangles

What then are the principal aesthetic factors * involved in polygonal form? Once these have been determined, we shall be able to define the corresponding elements of order in O, and the complexity C, and thus arrive at an appropriate aesthetic measure. We shall begin with the simplest class of polygons, namely the triangles.

Now triangles are usually classified as being either isosceles or scalene. Clearly the isosceles triangles, inclusive of the equilateral triangle, are more interesting from the aesthetic point of view. The usual orientation of an isosceles triangle is one in which the two equal sides are inclined at the same angle to the vertical, while the triangle rests on the third side. This is the case for each of the triangles (a), (b), (c) of the adjoining figure.

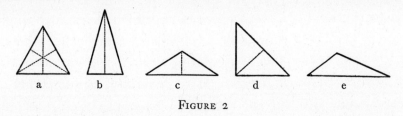

Figure 2

However, if these triangles are inverted, the equal sides will again be inclined at the same angle to the vertical. It is readily verified that this reversed orientation is also satisfactory. With these two orientations only do we obtain symmetry about a vertical axis. This is obviously a *desideratum* of prime importance.

The observation of symmetry about a vertical axis occurs constantly in everyday experience. Let us recall, for example, how quickly we become aware of any variation from symmetry in the human face. The association of vertical symmetry is intuitive and pleasing.

If the isosceles triangle (b) be made to rest upon one of the two equal sides, there still remains the feeling that the triangle is in equilibrium, although the symmetry about the vertical axis is thereby destroyed. It will be observed furthermore that the symmetry about the inclined axis is scarcely noted by the eye. Thus the triangle in its new orientation makes

---

* The term 'aesthetic factor' will be used in a non-technical sense.

much the same impression as any scalene triangle which rests upon a horizontal side (compare with (e)). The indifference of the eye to such an inclined axis of symmetry is also evidenced by the isosceles right triangle (d) with one of its equal sides horizontal.

On the other hand, if the isosceles triangle (b) be given any orientation whatsoever other than the two with vertical symmetry and the third just considered, there is a definite and displeasing lack of equilibrium.

For the isosceles triangle (c), however, there are only two orientations in which it seems to have full equilibrium, namely the two with vertical symmetry. In fact if the triangle (c) be made to rest upon one of its equal sides, the center of area falls so far to left or right as to give rise to the feeling that the equilibrium is not complete. The 'center of area' of any polygon is defined as that point of support about which it balances in a horizontal plane. Of course the polygon is assumed to be of uniform surface-density.

The association of equilibrium is also developed in our everyday experience.

Among the various shapes of isosceles triangles, the equilateral triangle (a) stands out as possessing peculiar interest because of its rotational symmetry. If such an equilateral triangle be set in a position which is not symmetrical about a vertical axis, all the pleasure in this symmetric quality disappears. However, once the favorable orientation is taken, the rotational symmetry is appreciated, largely by means of the three axes of symmetry. In the equilateral triangle the center of symmetry (and area) is the point of intersection of the three axes of symmetry, and the angle of rotation is 120° or one third of a complete revolution.

Associative reference to rotational symmetry often occurs in everyday experience. The form of the circle may perhaps be regarded as inducing this reference most completely.

Among the isosceles triangles which are not equilateral, there seems to be little to choose in respect to aesthetic merit. It does not appear to be a matter of importance whether the angle between the two equal sides is acute as in (b), obtuse as in (c), or a right angle. Of course when such a triangle is used in conjunction with other geometrical forms, this is no longer necessarily the case.

POLYGONAL FORMS

5. SCALENE TRIANGLES

The scalene triangles are readily disposed of. The best position is one in which the triangle rests upon a horizontal side long enough for the triangle to be in complete equilibrium. The right triangle with vertical and horizontal sides is obviously the best among the scalene triangles (note the triangle (d)). From our point of view this is because an unfavorable factor enters into the general scalene triangle of type (e) due to the presence of *three* unrelated directions.

6. CONCLUSIONS CONCERNING TRIANGULAR FORMS

Thus the various types of triangles in a vertical plane can be grouped in the following five classes according to order of aesthetic value: (1) the equilateral triangle with vertical axis of symmetry; (2) the isosceles triangle with vertical axis of symmetry; (3) the right triangle with vertical and horizontal sides; (4) any other triangle resting upon a sufficiently long horizontal side to ensure the feeling of complete equilibrium; (5) any triangle which lacks equilibrium. The triangles of the first two classes are definitely pleasing; those of the third class are perhaps to be considered indifferent in quality; and those of the fourth and fifth classes are displeasing. Since it is a natural requirement that the best orientation of any triangle be selected, the fourth class will contain all of the scalene triangles without a right angle, and the fifth class will scarcely enter into consideration.

It has been tacitly assumed in the above analysis of triangular form that no side of the triangle is extremely small in comparison to the other two sides, and that no angle is very small or very near to 180°. These are obvious prerequisites if the triangle is to be characteristic. If they are not satisfied, the triangle approximates in form to a straight line and the effect of ambiguity is definitely disagreeable. Likewise, when the triangle is very nearly but not quite isosceles, or very nearly but not quite equilateral, there is produced a feeling of ambiguity.

We are now in a position to list the aesthetic factors that have been so far encountered: vertical symmetry (+), symmetry about an inclined axis (o), equilibrium (+), rotational symmetry (+), diversity of directions (−), small side (−), small angle or angle nearly 180° (−), other ambiguity (−).

Here and later we use the symbol (+) to indicate that the corresponding factor is positive (that is, increases aesthetic value), the symbol (o) to indicate that it is without substantial effect, and the symbol (−) to indicate that it is negative (that is, diminishes aesthetic value). It is to be observed that, among the three positive factors, that of equilibrium is regarded as having a negative aspect also, arising from lack of equilibrium.

7. Plato's Favorite Triangle

The classification of the various forms of triangles given above takes only formal aesthetic factors into account. How completely such a classi-

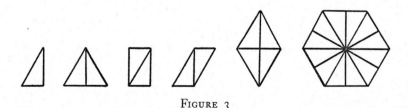

Figure 3

fication can be upset by the introduction of fortuitous connotative factors is easily illustrated.

Plato in the *Timaeus* says: * "Now, the one which we maintain to be the most beautiful of all the many figures of triangles (and we need not speak of the others) is that of which the double forms a third equilateral triangle." The context makes perfectly clear in what sense his statement is to be interpreted: If one judges the beauty of a triangle by its power to furnish other interesting geometrical figures by combination, there is no other triangle comparable with this favorite triangle of Plato. For out of it can be built (see Figure 3) the equilateral triangle, rectangle, parallelogram, diamond, and regular hexagon among polygons, as well as three of the five regular solids. This power in combination was peculiarly significant to Plato, who valued it for purposes of cosmological speculation. It was on such a mystical view that he based his aesthetic preference for this particular triangle.

* Translation by Jowett.

POLYGONAL FORMS

8. THE SCALENE TRIANGLE IN JAPANESE ART

It is well known that the Japanese prefer to use asymmetric form rather than the too purely symmetric. Indeed in all art, whether Eastern or Western, irrelevant symmetry is tiresome.

In particular it has been said that composition in Japanese painting is based upon the scalene triangle. Is this fact in agreement with the classification effected above, which concedes aesthetic superiority to the isosceles and in particular to the equilateral triangle? The answer seems to be plain: When used as an element of composition in painting, the isosceles triangle may introduce an adventitious element of symmetry. But, in the more elementary question of triangular form *per se*, the equilateral and isosceles triangles are superior to the scalene triangle.

Recently while in Japan I was fortunate enough to be able to ask one of the greatest Japanese painters, Takeuchi Seiho, if this were not the case, and he told me that the same opinion would doubtless be held in Japan.

9. SYMMETRIC QUADRILATERALS (FIRST TYPE)

Let us turn next to the consideration of the form of quadrilaterals, and let us examine those first in which there is symmetry about a vertical

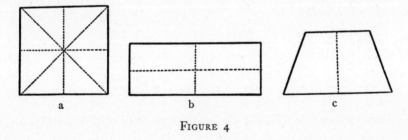

FIGURE 4

axis. There are two types. In the first, at least one side of the quadrilateral intersects the axis of symmetry; evidently such a side must be perpendicular to the axis of symmetry. Furthermore there must then be a second opposite side which is also perpendicular to the axis. Thus the general possibility is that of a symmetric trapezoid given by (c) of the figure above. This trapezoid may, however, take the form of a rectangle or square, illustrated by (b) and (a) respectively. It is to be observed that the

AESTHETIC MEASURE

rectangle possesses a further horizontal axis of symmetry, while the square possesses not only horizontal and vertical axes of symmetry but also two axes inclined at 45° to the horizontal direction. Likewise both rectangle and square have rotational symmetry, the angles of rotation being 180° and 90° respectively.

Corresponding to the degree of symmetry involved, we should expect to find the square to be the best in form, the rectangle excellent, and both superior in aesthetic quality to the symmetrical trapezoid. Such a relative rating coincides, I believe, with the facts.

It has been claimed sometimes that the rectangle is a form superior to the square, and even that certain rectangles such as the 'Golden Rectangle' surpass all others. We shall consider later (sections 14, 15) in what sense, if any, such assertions can be valid.

10. Symmetric Quadrilaterals (Second Type)

In the second type of symmetry about the vertical axis, the quadrilateral has two of its vertices on the axis of symmetry, but none of the

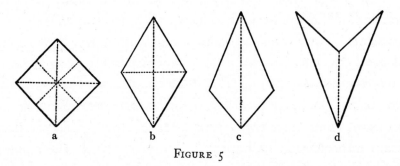

FIGURE 5

sides intersect the axis. Here the general possibility is indicated by (c) and (d) of the next figure (Figure 5) in which the deltoid (c) is convex, while (d) is re-entrant. The first of these may, however, take the form of an equilateral quadrilateral or diamond as in (b), or even of the square (a) with sides inclined at 45° to the horizontal direction.

Of the two general cases represented by the quadrilaterals (c) and (d) it is clear that the convex type (c) is definitely superior to the alternative re-entrant quadrilateral (d). The latter quadrilateral suggests a triangle from which a triangular niche has been removed.

POLYGONAL FORMS

It is not the mere fact that the quadrilateral is re-entrant which is decisively unfavorable. Consider, for example, the ordinary six-pointed star having the same outline as the familiar hexagram. This star is evidently highly pleasing in form, and yet is re-entrant. It will be noted, however, that every side, although of re-entrant type, is supported by another side which lies in the same straight line, while this is not true of the re-entrant sides of the quadrilateral above.

A re-entrant side will be termed 'supported' or 'unsupported' according as another side of the polygon does or does not lie in the same straight line.

In the comparison of quadrilaterals of types (a)–(d) we find, as we should expect, that the square (a) and the diamond (b) in the orientations indicated are markedly superior to the quadrilaterals (c) and (d) already discussed. However, it seems difficult to say whether or not the square so situated is better in form than the diamond, despite the fact that, on the score of symmetry alone, the square holds higher rank.

As far as I can analyze my own impressions, I am led to the following explanation of this aesthetic uncertainty: For me and many other persons the orientation of the square with sides vertical and horizontal is superior to the orientation with inclined sides. This superiority agrees with the theory of aesthetic measure, according to which the square in horizontal position has the highest rating of all polygonal forms ($M = 1.50$), while the square in the inclined position, together with the rectangle in horizontal position, come next ($M = 1.25$). Hence there arises a feeling of 'unnecessary imperfection' (Chapter I, section 11) when the square is in the inclined orientation, just because it would be *so easy* to alter it for the better. As soon as this association, which is really irrelevant, is abstracted from, the inclined square (a) will be found, I believe, to be superior to the diamond (b).

11. Symmetric Quadrilaterals (Third Type)

There remain for discussion those quadrilaterals which are not symmetric about an axis. Here, as in the case of the scalene triangle, attention may be limited to cases in which the quadrilateral rests on a sufficiently long horizontal side so that it appears to be in complete equilibrium.

It is readily seen that the only quadrilaterals of this type which possess rotational symmetry are the parallelograms, illustrated by (a) in the adjoining figure. Evidently the angle of rotation for a parallelogram is 180°, so that it possesses central symmetry.

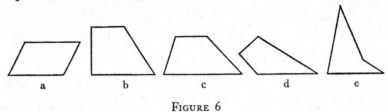

FIGURE 6

The parallelogram is the simplest polygon possessing rotational symmetry but not symmetry about the axis. It is more satisfactory in form than the remaining asymmetric quadrilaterals, just because of its symmetric character.

12. Asymmetric Quadrilaterals

Among the asymmetric quadrilaterals, represented by (b), (c), (d), (e) in the figure, the right-angle and general trapezoids with two sides parallel, as in (b) and (c), stand first.

The associative reference to 'parallelism' occurs frequently in everyday experience. The factor of parallelism is closely correlated with that of 'diversity of directions,' already listed. For, the more sides are parallel, the less will be the diversity of directions.

After the trapezoids in aesthetic quality follow the general convex quadrilateral (d), and finally the re-entrant case (e). It will be observed that the presence of an isolated right angle or of two equal sides, as in (d), is without noticeable influence.

13. Conclusions concerning Quadrilateral Forms

We have now examined the types of quadrilateral forms and have arranged those of each type in order of aesthetic merit. It remains to compare briefly those of different types.

The usual order of preference appears to be in the following groups of diminishing aesthetic value: the square; the rectangle; the diamond; the symmetric trapezoid, the deltoid, and the parallelogram; the re-entrant quadrilateral symmetric about an axis; the right-angle trapezoid; the remaining convex quadrilaterals and re-entrant quadrilaterals without symmetry. Of these, the last group is definitely unsatisfactory. It is assumed here that the quadrilaterals are placed in the best vertical orientation. This relative arrangement is the same as that assigned by the theory of the present chapter. We shall not attempt at this juncture to compare quadrilaterals and triangles.

Thus there are two kinds of aesthetic factors brought to light by our examination of quadrilaterals. The first is of negative type and is occasioned by the fact that the quadrilateral is re-entrant. Further insight into the nature of this factor will be obtained as we proceed. The second factor is connected with the parallelism of sides. As we have observed, it is more convenient to regard this second factor in its negative aspect, when it appears as that of diversity of directions.

It has been seen also that the mere equality of sides is an indifferent factor for quadrilaterals, as it is of course for polygons having more than four sides. This conclusion stands in sharp distinction with that for the triangle. The reason for the difference lies in the fact that only for the triangle does equality of two sides ensure symmetry.

14. The 'Golden Rectangle' and Others

The so-called Golden Section of a linear segment is that which divides it in two segments in such a way that the longer segment is the mean proportional between the shorter segment and the whole segment.

The mathematician Luca Paciolo had claimed long ago* the central aesthetic importance of the proportion of the Golden Section. Within the last seventy-five years particularly this doctrine has been the subject of further speculation and of interesting experimental investigations. In particular the 'Golden Rectangle' (as we shall call it) whose sides are in the ratio of the Golden Section has attracted especial attention. This special rectangle with a ratio of 1.618 . . . , and so very nearly 8 to 5, is obviously agreeable to the eye. Furthermore it has the very interesting geometric property that if a square on the shorter side be removed, a smaller Golden Rectangle remains. Is it not then perhaps true that

* *De divina proportione*, Venice (1509).

AESTHETIC MEASURE

there resides in the Golden Rectangle some occult beauty which makes it superior to all other rectangular forms?

The psychologist Fechner * conducted well known experiments to ascertain if possible the most satisfactory rectangular shape, inclusive of the square. His results may be briefly summarized as follows: The square and, more especially, rectangles having dimensions approximating those of the Golden Rectangle, were generally considered to be the best.

However, Fechner used rectangular picture frames and a variety of other rectangular objects in his experiments. Such a frame is determined in its dimensions by the nature of the picture. In other cases also the dimensions are determined by similar considerations. To this extent his

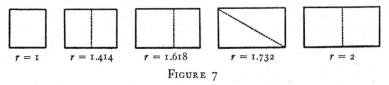

$r = 1$ $r = 1.414$ $r = 1.618$ $r = 1.732$ $r = 2$

FIGURE 7

observations are irrelevant to the question of rectangular forms considered in isolation.

In thinking of the Golden Rectangle, it is well to keep in mind the special rectangle favored by Plato, made of the two halves of an equilateral triangle. For it, the characteristic ratio is 1.732. . . . It is also well to keep in mind the rectangle with ratio 1.414 . . . , which may be divided in two equal rectangles of the same shape as the original rectangle by a line parallel to the two shorter sides. Here the ratio is nearly 7 to 5. Furthermore the rectangle made up of two squares, with ratio 2 to 1 is to be noted.

Thus we find five rectangles (we include the square) with simple geometric properties. These are represented above, and the ratio r of the longer side to the shorter is given in each case. We conclude then that the Golden Rectangle is in no wise different from others in the respect that it possesses a simple geometrical property.

15. COMPARISON OF RECTANGULAR FORMS

If now we regard these and other rectangular forms as embodied in isolated vertical tiles, it will be found in the first place that the square tile

* See his *Vorschule der Ästhetik*, Leipzig (1876), in particular chapter 19.

is the most effective. This fact agrees with the observation that the square is more frequently used in this manner than any other single rectangular shape. It will also be found that other rectangular forms are highly pleasing and not much to be distinguished from one another in aesthetic quality excepting as follows: a rectangular form nearly but not quite a square is disagreeable because of the effect of ambiguity (Chapter I, section 11); a rectangular form in which the ratio of the longer to the shorter side is too great suggests the segment of a line, and this effect of ambiguity is also disagreeable.

Moreover a rectangular form with a ratio as much as 2 to 1 is not well adapted to fill the circular field of effective vision. In consequence, such forms are not suitable in many cases, as for instance that of picture frames. But for a rectangular tile, such as we are considering, this is certainly not the case, since tiles with a considerably greater ratio are very pleasing and are used frequently for decorative purposes.

To avoid all suspicion of either ambiguity or lack of utility, we must therefore restrict ourselves further to rectangles whose characteristic ratio is plainly between 1 and 2. In consequence if we desire to choose rectangular forms which completely avoid undesirable factors, we are inevitably led to forms not very far from that of the Golden Rectangle, but among which are others like that with ratio $r = 1.414$ ... in the figure above. All such rectangular forms are both pleasing and useful.

These conclusions are in substantial agreement with the general theory of the present chapter, which accords a leading position to the square ($M = 1.50$) and all unambiguous rectangular forms ($M = 1.25$), but does not take account of the usefulness of rectangular forms. It is to be observed that usefulness corresponds to a connotative factor entirely outside of the scope of the theory.

Lipps has expressed himself to much the same effect as follows:*

It may now be looked upon as generally conceded ... that the ratio of the Golden Section, generally and in this case [of the Golden Rectangle], is entirely without aesthetic significance in itself, and that the presence of this numerical ratio is nowhere the basis of any pleasant quality. ...

In this way the question arises as to whence comes the indubitable special agreeableness of rectangles approximating that of the Golden Section. ...

* *Ästhetik*, vol. 1, pp. 66–67, my translation.

The rectangles in question are just those in which the smaller dimension is decisively subordinated to the greater. . . .

It has already been indicated above why the rectangle which approaches the square too closely pleases little. We term it awkward because of its ambiguity. On the other hand the rectangle in which one dimension falls too much behind the other . . . seems insufficient, thin, attenuated.

The experimental results of Fechner may also have been influenced by the fact that numerous persons, through their acquaintance with and liking for Greek art, or otherwise, have come to individualize and identify the Golden Rectangle. For them a connotative association of purely accidental character would be established in its favor. If a number of his experimental subjects were of this sort, Fechner's experimental results are easily explained.

16. Use of Rectangular Forms in Composition

Up to this point we have considered rectangular forms in isolation. It is interesting to note some facts concerning their use in composition, which undoubtedly have some residual effect upon our general appreciation of them even in isolation.

For use in composition it is very important that the rectangles have an infinitude of shapes, dependent on the arbitrary ratio of the sides, whereas the square has a single definite shape. Hence the rectangles provide a much more flexible instrument than does the square. It is, for example, obvious that the square shape is not in general suitable as a frame for a picture. This superior usefulness of the rectangle may establish in the long run a positive connotative factor in its favor.

Furthermore, in many of its uses, such as that of a picture frame, any obvious numerical ratio of dimensions such as 1 to 1 or 2 to 1 is to be avoided because it is often desirable that the rectangle be a purely neutral accessory, not producing irrelevant associations.

Finally it is to be observed that although the special forms of rectangles, like those mentioned above, have no especial significance when used in isolation, this is no longer true when they appear as elements in composition. For instance, the arrangement of two adjoining rectangular windows with $r = 1.414 \ldots$ so as to form a single rectangle of the same shape might be decidedly pleasing architecturally, because the same shape is

POLYGONAL FORMS

discovered in an unexpected aspect. Similarly, three adjoining windows, the central one being a square and the outer ones equal Golden Rectangles with the shorter sides horizontal, might prove very pleasing.

17. The Forms of Five- and Six-sided Polygons

Our survey of triangles and quadrilaterals has brought to light a number of the essential aesthetic factors which operate in more general cases. It would be tedious to continue with our analysis in full detail. If we did so, the facts for five- and six-sided polygons would be found to be on the whole similar to those already noted. We shall be content, therefore, to mention those factors which are not illustrated by the polygons of three or

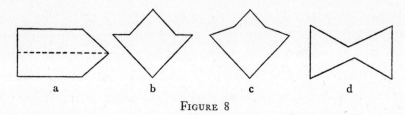

Figure 8

four sides, and then to pass on to such as are illustrated only by still more complicated polygons.

To begin with, let us recall that symmetry about an inclined axis has been observed to be of little or no aesthetic significance by itself. It is true that when there is symmetry about the vertical axis also, the matter is not so clear, but in that event there will be rotational symmetry as well. In consequence such symmetry about an inclined axis is to be looked upon as arising from the combined vertical symmetry and rotational symmetry, and so as logically dependent upon them. Hence symmetry about an inclined axis need not be considered in this case as a separate aesthetic factor.

What is the importance of symmetry about a horizontal axis when there is no vertical symmetry? This case is illustrated most simply by the pentagonal polygon (a) in Figure 8. In the first place it is clear that the symmetry about the horizontal axis is much more easily appraised by the eye than in any other direction except the vertical. Notwithstanding this fact, however, the symmetry about the horizontal axis is not enjoyed.

AESTHETIC MEASURE

A second factor already briefly alluded to is effectively isolated by a comparison of the two re-entrant hexagonal polygons (b) and (c). These are both of the same general type, but only in the first case (b) do two of the four re-entrant sides lie in a straight line and so support one another. It is obvious that (b) is notably superior to (c) just on this account. In general, then, we may expect unsupported re-entrant sides to operate unfavorably and so to correspond to a factor of negative type.

The psychological explanation of this situation appears to be as follows: in general the association of re-entrance is not a pleasant one; but if, when the eye follows a re-entrant side, another side is discovered in the same straight line there is a compensating feeling of satisfaction.

A new form of ambiguity is illustrated by the hexagon (d), in which two parallel sides are found nearly in the same straight line.

18. More Complicated Forms

The 90 polygons listed in the opposite Plates II–VII in order of decreasing aesthetic measure present graphically some of the principal types of polygons. Examination of these polygons yields a few further aesthetic factors of importance.

In the first place there is the obvious increasing complexity itself, which, beyond a certain point, is found to be tiresome. Furthermore it is evident that the convex polygon with a large number of sides is more likely to be pleasing than the re-entrant one, particularly if the latter contains a diversity of niches.

By a 'niche' of a polygon is meant any outer area lying within the minimum enclosing convex polygon.

Another important factor becomes obvious when the polygon is closely related to some uniform network of horizontal and vertical lines. The relationship may be direct, as in the case of the Greek cross (see (a) in the following figure); or it may be indirect, in that the polygon is directly related to a uniform diamond network with its sides equally inclined to the vertical (see (b)), while this network in turn suggests a uniform horizontal-vertical network.

Evidently the aesthetic factor of close relationship to a uniform horizontal-vertical or diamond network enhances greatly the aesthetic value

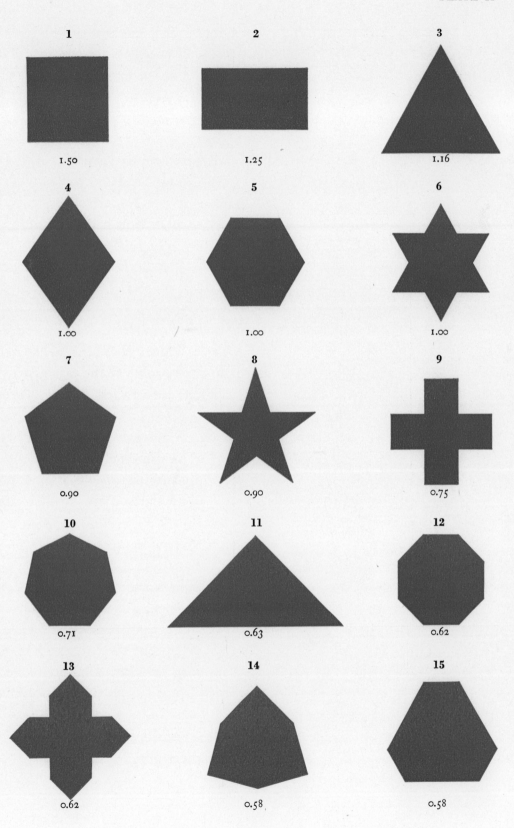

Aesthetic Measures of 90 Polygonal Forms, Nos. 1–15

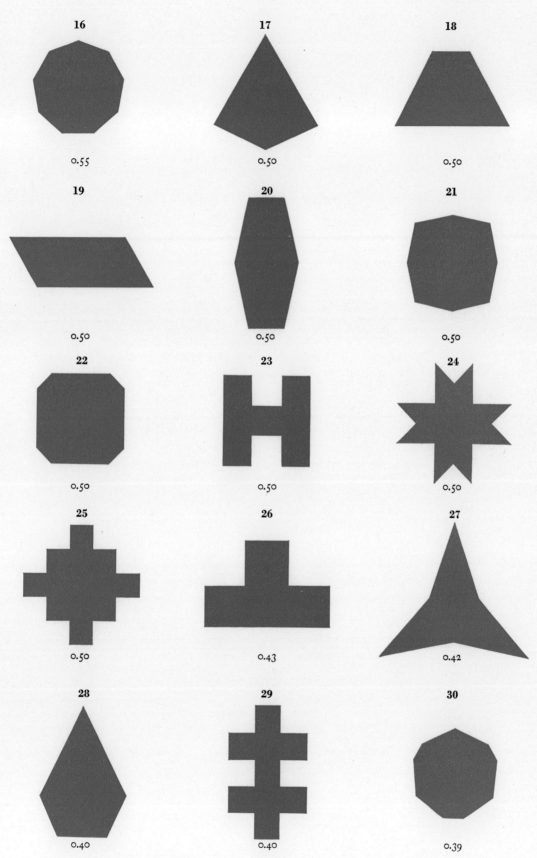

AESTHETIC MEASURES OF 90 POLYGONAL FORMS, NOS. 16–30

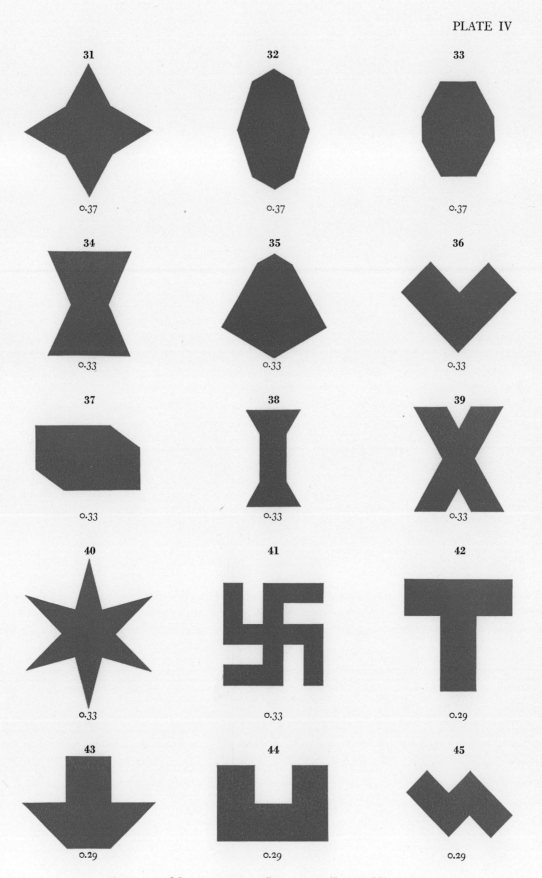

AESTHETIC MEASURES OF 90 POLYGONAL FORMS, NOS. 31–45

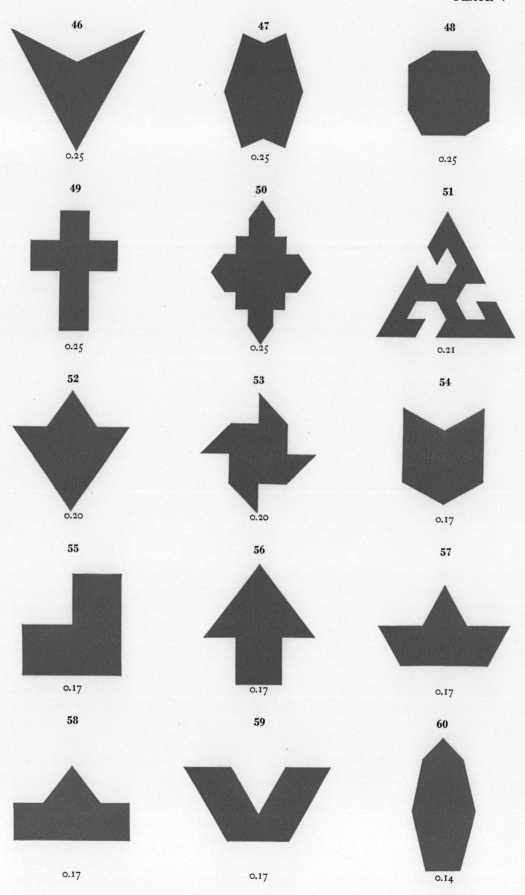

Aesthetic Measures of 90 Polygonal Forms, Nos. 46–60

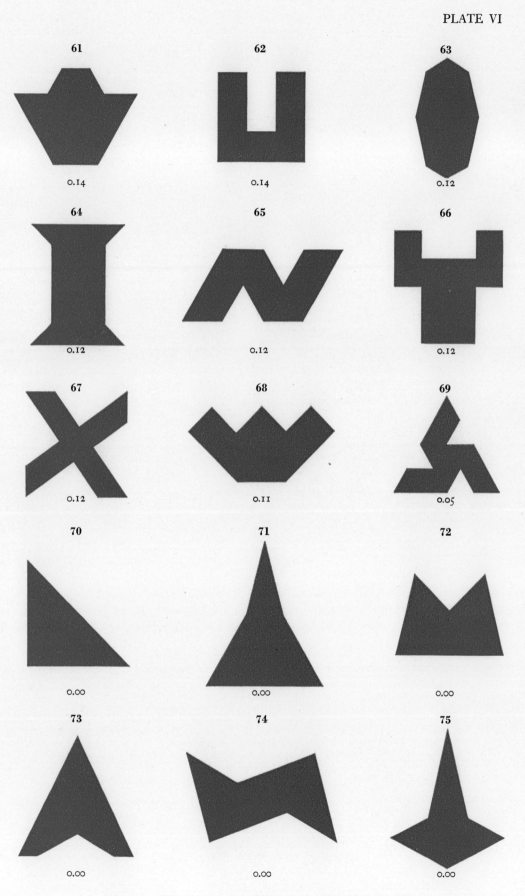

Aesthetic Measures of 90 Polygonal Forms, Nos. 61–75

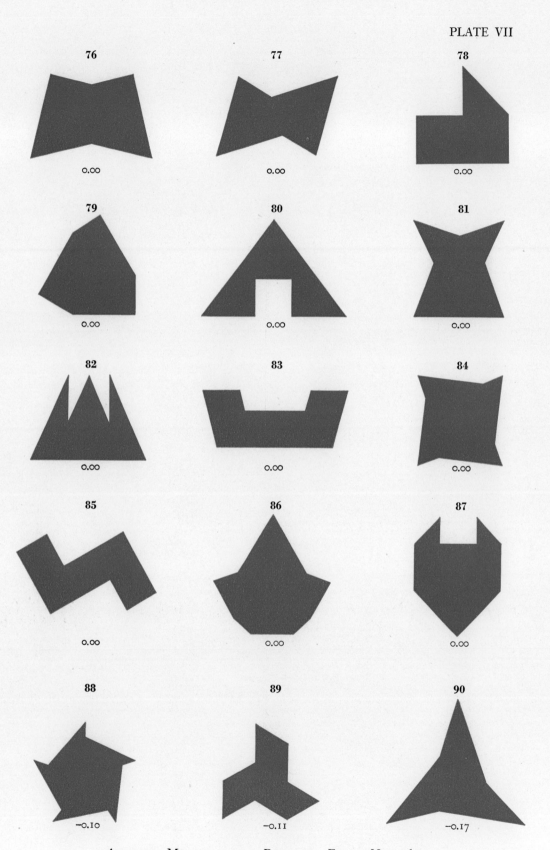

Aesthetic Measures of 90 Polygonal Forms, Nos. 76–90

POLYGONAL FORMS

of many of the polygons listed, e.g. the square (No. 1), the rectangle (No. 2), the diamond (No. 4), the six-pointed star (No. 6), the Greek cross (No. 9), the swastika (No. 41), etc. In the case of the square, rectangle, and diamond, the validity of this association may seem debatable, but so constantly do we observe all these forms in a uniform network that they seem always to suggest such a network; however, direct relation to a diamond network possesses less value than direct relation to a horizontal-vertical network. The associational basis of this factor in everyday experience is obvious: Systems of lines placed in the regular

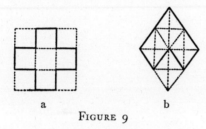

a   b

FIGURE 9

array of a uniform network are constantly met with, and their relationship to one another is intuitively appreciated.

The examination of more complicated polygons shows also that some kind of symmetry is required if their form is to be at all attractive. When this requirement is not met, relationship to a horizontal-vertical network, for instance, will not offset the deficiency. Evidently a further actual aesthetic factor in many cases is some accidental connotation, such as is present in the case of crosses and the swastika. Our theory leaves such connotative factors out of account.

19. On the Structure of the Aesthetic Formula

According to the general theory proposed in the first chapter, we seek an aesthetic formula of the type $M = O/C$ where M is the aesthetic measure, O is the order, and C is the complexity. In the case of polygonal form, O will be separated into five elements of order:

$$O = V + E + R + HV - F.$$

The aesthetic factors encountered above are correlated in the following way with C and the five elements of order which make up O:

C: complexity ($-$),
V: vertical symmetry ($+$),
E: equilibrium ($+$),
R: rotational symmetry ($+$),
HV: relation to a horizontal-vertical network ($+$),
F: unsatisfactory form ($-$) involving some of the following factors: too small distances from vertices to vertices or to sides, or between parallel sides; angles too near 0° or 180°; other ambiguities; unsupported re-entrant sides; diversity of niches; diversity of directions; lack of symmetry.

It will be observed that the term F is an *omnium gatherum* for the negative aesthetic factors of unsatisfactory form. The various indifferent factors of type (o) play no part of course. Among these are equality of sides, and inclined or horizontal symmetry.

In the course of the technical evaluation of C, V, E, R, HV, F, and so of M, to which we now proceed, a simple mathematical concept, namely that of the group of motions of the given polygon, will be introduced (section 23). This concept will prove to be a useful adjunct.

20. THE COMPLEXITY C

The complexity C of a polygon will be defined as the number of indefinitely extended straight lines which contain all the sides of the polygon. Thus for a quadrilateral the complexity is evidently 4; for the Greek cross the complexity is 8, although the number of sides in the ordinary sense is 12; for the pinwheel figure shown in No. 53 the complexity is 10.

The psychological reasonableness of this empirical rule is evident: For convex polygons, and also for re-entrant polygons without any two sides in the same straight line, the complexity C is given by the number of sides. As the eye follows the contour of the polygon, the effort involved is proportional to this number. On the other hand, if there are several sides in one and the same straight line, the eye follows these in one motion. For example, in the case of the Greek or Roman cross, the eye might regard it as made up of two rectangles. These considerations suggest that the definition chosen for the complexity C is appropriate.

POLYGONAL FORMS

21. The Element V of Vertical Symmetry

The agreeable organization of the polygon which results from vertical symmetry is obvious to the eye. By long practice we have become accustomed to estimating symmetry of this sort immediately. The corresponding element of order V is particularly significant.

We shall give to V the value 1 when the polygon possesses symmetry about the vertical axis, and the value 0 in the contrary case.

In other words, the element V will be a unit element of order, and, since there exist various polygons of pleasing quality, such as the swastika, which do not have vertical symmetry, we shall assign the value 0 rather than a negative value to V when there is no such symmetry.

When there is vertical symmetry, a center of repose for the eye will be found upon the axis of symmetry. Furthermore the polygon possessing such symmetry is felt to be in equilibrium. A large proportion of the 90 polygons listed possess vertical symmetry, and it can be verified that such symmetry is favorably felt.

22. The Element E of Equilibrium

Let us consider the second element E of order, concerned with equilibrium. It has been previously observed that when the polygon has vertical symmetry or rests upon a sufficiently extended horizontal base, it is felt to be in complete equilibrium.

In order to specify the requirements for complete equilibrium, we may note first that it is optical equilibrium which is referred to rather than ordinary mechanical equilibrium. For example, the pinwheel polygon No. 53 is in (unstable) mechanical equilibrium, inasmuch as the center of area lies directly above the lowest point. Nevertheless it does not give the impression of optical equilibrium. In the case when the polygon is not symmetric about the vertical, the feeling of optical equilibrium is only induced if there is a horizontal base with the extreme points of support far enough apart so that the center of gravity lies well between the vertical lines through these extreme points.

We shall agree that there is complete optical equilibrium if the center of area lies not only between these two vertical lines, but at a distance from either of them exceeding one sixth that of the total horizontal breadth of the polygon. When this arbitrary condition is satisfied, as well as in the case of vertical symmetry, we shall give E the value 1. If the polygon does not satisfy this condition but is in equilibrium in the ordinary mechanical sense, we shall take E to be 0. Otherwise we shall take E to be -1, inasmuch as the lack of equilibrium is then definitely objectionable.

In all of the listed polygons the selected orientation gives at least mechanical equilibrium, although in one case, No. 85, an equally favorable orientation exists lacking equilibrium ($E = -1$), in which the sides make an angle of 45° with the vertical.

23. The Group of Motions of a Polygon

In the case when the polygon possesses rotational symmetry, it has been observed (section 3) that there is a least angle of rotation $360°/q$. If the corresponding rotation be effected, the polygon will be returned as a whole to its initial position. If it be repeated q times, a complete revolution will be effected, so that every point returns to where it was at the outset.

There is a certain fundamental similarity between such rotational symmetry and the symmetry about an axis. In order to make this clear, let us recall that if a polygon be rotated through 180° about an axis of symmetry, it will return to its initial position. Thus the test for both kinds of symmetry is that a certain rotation restores the polygon as a whole to its initial position. In the case of rotational symmetry, the axis of rotation is perpendicular to the plane of the polygon at its center of area; while in the case of axial symmetry the axis of rotation is the axis of symmetry, and the rotation is a half revolution.

The collection of all these motions of rotation leaving a given polygon in the same position constitutes the 'group of motions' of the polygon. For reasons of convenience it is desirable to admit, as a conventional motion of rotation, the rotation about an arbitrary axis through an angle of 0°, which moves no point.

If A denotes one such rotation, and B the same rotation or any other, then the combination of the rotation A with the subsequent rotation B may be denoted by AB, and returns the polygon to its initial position.

POLYGONAL FORMS

Hence the compound operation thus effected must be equivalent to a single rotation C of the group; that is, we may write, in brief mathematical symbolism, $AB = C$.

As a simple example let us consider a square. The 'group of motions' contains the following eight rotations: four rotations of 0°, 90°, 180°, 270° in its plane about the center of area; four rotations of 180° about the two diagonals and their two bisectors.

Suppose now that we follow the motion of rotation about one diagonal by a rotation about the other. The resultant effect is to transfer each point to the centrally symmetric point, just as if the polygon had been rotated in its plane through 180°. In other words the compound rotation formed from these two rotations of the group is equivalent to another rotation of the group. This fact illustrates the fundamental principle embodied in the symbolic equation $AB = C$.

The figures in the plane of the given polygon which arise from one such figure when all possible rotations of the group are applied to it will be said to be 'of the same type' as the given figure. In more mathematical terms all such figures are 'conjugate' under the given group.

For example, in the case of the square the four vertices, and also the four sides, are of the same type; vertical and horizontal directions are of the same type; the diagonals are of the same type. In the case of the rectangle all four vertices but only pairs of opposite sides are of the same type; similarly, vertical and horizontal directions are not of the same type, although the diagonals are.

Figures 'of the same type' are merely *corresponding* figures in the intuitive sense of the term.

The groups of motions of polygonal forms are of three possible species: (1) the groups of the regular polygons of q sides,* in which there is both axial and rotational symmetry; (2) the group of the isosceles triangle, in which there is only symmetry with respect to a single axis; (3) groups like those of the parallelogram ($q = 2$) and the swastika ($q = 4$) in which there is rotational symmetry with an angle of rotation $360°/q$ but no symmetry about an axis.

* Inclusive of the case of a 'regular polygon of two sides' formed by a single line. The group in this case is the same as that of a rectangle.

24. The Element R of Rotational Symmetry

In dealing with the element R of rotational symmetry, we are guided by certain obvious visual facts. The simplest type of rotational symmetry is that of central symmetry in which the least angle of rotation is 180°. The parallelogram illustrates this possibility. It is clear that central symmetry is at once appreciated by the eye.

In the case of central symmetry any line through the center of area divides the polygon into two figures 'of the same type.' For example, a diagonal of the parallelogram divides it into two triangles of the same type. On this account any half of such a polygon determines the other half. Likewise in the case of axial symmetry, the half of the polygon on one side of the axis determines the half on the other side, and these two halves are also of the same type. In other words, central symmetry requires the same extent of organization within the polygon as does axial symmetry.

For this reason we take the element R to be 1 in the case when there is only central symmetry, just as we take V to be 1 in the case of axial (vertical) symmetry.

Let us turn next to the case where the group of motions is that of a regular polygon. Obviously the only effective orientations are those in which an axis of symmetry lies along the vertical direction; the rotational symmetry is appreciated more for larger values of q.

Here the polygon can be broken up into q partial symmetric polygons situated symmetrically around the center of area. By analogy with the case of central symmetry, it is therefore natural to assume that R varies in proportion to q; for, any one of these q component parts determines all of the others. Since when q is 2, R is 1, we are in general led to define R as $q/2$.

However, the element of rotational symmetry is only effective up to a certain point, after which there is no further increase. When q is 6 or exceeds 6, the circle circumscribed about the polygon is very clearly suggested and the impression of rotational symmetry becomes complete. For these reasons, in the case of vertical and rotational symmetry combined we define the element R as $q/2$ for q not greater than 6, and as 3 for larger values of q.

POLYGONAL FORMS

There is a re-entrant case in which the element R is felt almost equally favorably, despite a lack of vertical symmetry, namely when the minimum convex polygon enclosing the given polygon is symmetric about a vertical axis, and none of its vertices abut on the niches. Polygons Nos. 41, 51, and 69 illustrate this case; here the enclosing convex polygon, with its q axes of symmetry, is so strongly outlined as to suggest vividly the rotational element, and we define R as in the preceding case.

On the other hand, even though the minimum enclosing convex polygon is symmetric about a vertical axis, the same effect is not felt if its vertices abut on the niches of the given polygon. This possibility is illustrated by polygons Nos. 53, 67, 84, 88, and 90. In explanation of this difference in effect, it may be observed that the axes of symmetry of the enclosing polygon are hardly felt as such under these circumstances.

In such cases and in others when the enclosing polygon is not axially symmetric, there is central symmetry when q is even. Such symmetry is appreciated immediately, though the rotational symmetry as such plays a negligible rôle. Accordingly we take R to be equal to 1 here. Polygons Nos. 48, 53, and 67 illustrate this possibility.

On the other hand, when q is odd the effect is less favorable still. Nos. 79 and 88, 89, 90 illustrate this situation for convex and re-entrant polygons respectively. Even in the convex type, most persons will scarcely be aware of the rotational symmetry or will find it to be disagreeable. Thus in this last case, as well as in any case when there is no rotational symmetry, we are led to take R to be 0.

25. THE ELEMENT HV OF RELATION TO A HORIZONTAL-VERTICAL NETWORK

As has previously been noted, in many polygons of the list there is evidently a close relationship of the given polygon to a uniform horizontal-vertical network, and this relationship is decidedly pleasing.

The corresponding element HV in O is connected with certain motions of the plane in much the same way as the element V is connected with a motion of rotation about a vertical axis, and the element R with a motion of rotation about a center. In fact such a uniform network evidently returns as a whole to its initial position, when certain translatory motions of the plane are made. Such motions will take the polygon to a new position related to the same network. In an incomplete way, then, the element HV is connected with motions of the plane just as the elements V and R are.

The most favorable case is that in which the polygon has all of its sides lying along the lines of a uniform network of horizontal and vertical lines, in such wise that these lines completely fill out a rectangular portion of the network. In this case only do we take HV to be 2. Nos. 1, 2, 9, 23, 25, 26, 29, 41, 44, and 55 illustrate this possibility.

The choice of 2 as the corresponding index of HV is suggested by the fact that there are essentially *two* kinds of independent translatory motions which return the network to its original position, namely a translation to the right or left, and a translation up or down. Any other translation may be regarded as derivable by combination from these two alone.

Another similar case is that in which the sides of the polygon all lie upon the lines of a uniform network formed by two sets of parallel lines equally inclined to the vertical, and fill out a diamond-shaped portion of the network (see Nos. 4, 36, and 45). But the effect here is less favorable, so that we take HV to be 1.

It becomes necessary at this stage to assign an arbitrary index in all cases. From the purely geometrical point of view, the degree of coincidence of a polygon with such a horizontal-vertical or diamond network may range from the case of maximum coincidence, as specified above, to the case of practically no coincidence, through a series of intermediate degrees. It is thus suggested that a corresponding graded index may be required. It is found, however, that as soon as there is a slight deviation from complete coincidence, the pleasantness of the effect diminishes markedly, and for further deviation vanishes entirely.

Hence we shall select the following empirical rule: HV is to be 1 if the polygon fills out a rectangular portion of a horizontal-vertical network, save for the following exceptions: one line of the polygon and the other lines of the same type (see section 23) may fall along diagonals of the rectangular portion or of adjoining rectangles of the network; one vertical and one horizontal line of this portion, as well as other lines of the same type, may not be occupied by a side of the polygon.

Illustrations of this case $HV = 1$ are furnished by the polygons Nos. 13, 42, 43, 49, 50, 53, 62, 66, and 78 of the list.

The element HV will also be defined to be 1 if the polygon fills out a diamond-shaped portion of a diamond network save for entirely analogous exceptions: one line of the polygon and the other lines of the same type may fall along diagonals of the diamond-shaped portion or of adjoining diamonds of the network; one line of this portion as well as the other lines of the same type may not be occupied by a side of the polygon. The polygons Nos. 5, 6, 24, and 68 are illustrative of this case.

Moreover when HV is 1 we shall demand that at least two lines of either set of the network are occupied by a side.

In all other cases whatsoever we shall take HV to be 0.

It is obvious that the above determination of indices for the element HV is largely arbitrary. Nevertheless it seems to correspond to the facts observed.

26. The Element F of Unsatisfactory Form

There remains to be treated the negative constituent F in O, which we have described as an *omnium gatherum* of the negative elements of order (section 19).

The case in which F is 0 corresponds to satisfactory form. Here the analysis made in the earlier sections suggests the following conditions: (1) the minimum distance from any vertex to any other vertex or side, or between parallel sides, is not to be too small — for definiteness we shall demand that it be not less than one tenth the maximum distance between points of the polygon; (2) the angle between two non-parallel sides is not to be too small — for definiteness let us say not less than 20°; (3) more generally, all other ambiguities of form are to be avoided — for definiteness let us demand that no shift of the vertices by less than one tenth their distance to the nearest vertex can introduce a further element of order in V, R, or HV; (4) there is to be no unsupported re-entrant side; (5) there is to be at most one type of niche; (6) there are to be at most two types of directions, provided that vertical and horizontal directions (when both occur) are counted together as one; (7) there is symmetry to the extent that V and R are not both 0.

The first three of these requirements eliminate ambiguity of form. Of them the first two deal with kinds of ambiguity explicitly mentioned above. The third requirement takes account of the kind of ambiguity which is found to arise, for example, when a triangle is nearly but not quite isosceles or equilateral.

The other requirements are also suggested by the analysis given above. In particular the fifth and sixth requirements are based on the fact that while one type of niche is compatible with satisfactory form, as in the Greek cross No. 9, and two types of directions are not excessive (provided vertical and horizontal directions are counted together as one), it is not possible to go further without impairment of satisfactory form.

If one and only one of the above conditions fails and that to a minimum extent (e.g. there is *one* type of unsupported re-entrant side), we take F to be 1. In all cases where there is more than a single violation of these conditions we take F to be 2.

For all of the polygons Nos. 1–22 inclusive, F is 0. No. 23 is the first polygon for which F is 1 because of one type of unsupported re-entrant side; No. 24 is the first polygon for which F is 1 because of two types of niches. The earliest polygon for which F is 1 because of diversity of directions is No. 60. The earliest for which F is 2 is No. 55.

27. Recapitulation of Definition of Aesthetic Measure

For purposes of convenient reference let us state concisely the above definition of the aesthetic measure of a polygon in vertical position:

The formula is
$$M = \frac{O}{C} = \frac{V + E + R + HV - F}{C}$$
with the following definitions:

$$C$$

C is the number of distinct straight lines containing at least one side of the polygon.

$$V$$

V is 1 or 0 according as the polygon is or is not symmetric about a vertical axis.

POLYGONAL FORMS

E

E is 1 whenever V is 1.

E is also 1 if the center of area K is situated directly above a point D within a horizontal line segment AB supporting the polygon from below in such wise that the lengths AD and BD are both more than 1/6 of the total horizontal breadth of the polygon.

E is 0 in any other case when K lies above a point of AB, even if A and B coincide.

E is -1 in the remaining cases.

R

R is the smaller of the numbers $q/2$ and 3 in the case of rotational symmetry, provided that the polygon has vertical symmetry or else that the minimum enclosing convex polygon has vertical symmetry and that the niches of the given polygon do not abut on the vertices of the enclosing polygon.

R is 1 in any other case when q is even (i.e. if there is central symmetry).

R is 0 in the remaining cases.

HV

HV is 2 only when the sides of the polygon lie upon the lines of a uniform horizontal-vertical network, and occupy all the lines of a rectangular portion of the network.

HV is 1 if these conditions are satisfied, with one or both of the following exceptions: one line and the others of this type may fall along diagonals of the rectangular portion or of adjoining rectangles of the network; one vertical line and one horizontal line of the portion, and the others of the same type, may not be occupied by a side. At least two vertical and two horizontal lines must be filled by the sides however.

HV is also 1 when the sides of the polygon lie upon the lines of a uniform network of two sets of parallel lines equally inclined to the vertical, and occupy all the lines of a diamond-shaped portion of the network, with the following possible exceptions: at most one line and the others of the same type may fall along diagonals of the diamond-shaped portion or of adjoining diamonds of the network; one line of the diamond-shaped portion and the others of its type may not be occupied by a side. At least two lines of either set of parallel lines in the network must, however, be occupied by the sides.

HV is 0 in all other cases.

$$F$$

F is 0 if the following conditions are satisfied: the minimum distance from any vertex to any other vertex or side or between parallel sides is at least 1/10 the maximum distance between points of the polygon; the angle between two non-parallel sides is not less than 20°; no shift of the vertices by less than 1/10 of the distance to the nearest vertex can introduce a new element of order V, R, or HV; there is no unsupported re-entrant side; there is at most one type of niche and two types of directions, provided that vertical and horizontal directions are counted together as one; V and R are not both 0.

F is 1 if these conditions are fulfilled with one exception and one only.

F is 2 in all other cases.

28. Application to 90 Polygons

The 90 polygons arranged in Plates II–VII in order of decreasing aesthetic measure according to the formula furnish in themselves a severe test of its approximate accuracy. If, upon scanning these polygons from the first to the last, the reader feels a gradual diminution in aesthetic quality,* the underlying theory may be regarded as justified. The following facts should be observed.

Many polygons have important connotative elements of order which will have an effect upon the aesthetic judgment, unless one abstracts from them explicitly. Thus the stars Nos. 6, 8, 40, the different crosses Nos. 9, 13, 29, 49, and the swastika No. 41, have their aesthetic value definitely enhanced by the corresponding connotations. The importance of the positive connotative elements of order in these cases becomes obvious if one inverts the Roman cross No. 49, and observes how its aesthetic effect is thereby impaired.

Likewise certain other polygons have important negative connotative elements of order. Thus polygons Nos. 23, 38, and 39 strongly suggest

---

* Those with the same aesthetic measure are to be grouped together of course.

the respective letters H, I, and X; this association operates to diminish their aesthetic value. Similarly the association of No. 37 with the outline of a rectangular box is displeasing. The polygon No. 56, of a shape which suggests either a pine tree or an arrowhead, will be increased or diminished in value perhaps, according as the first or second of these associations is uppermost.

Other connotations of a geometrical type ought to be noted. Thus the swastika, No. 41, suggests strongly two broken lines. A polygon of many small sides inscribed in a regular curve (see Nos. 10, 12, and 16) will suggest the curve so strongly as to exclude the consideration of the polygon as such.

In testing the validity of the formula as applied to such a set of polygons, all these connotations must be borne in mind.

Secondly, it is to be recalled that no special attempt is made here to classify indifferent polygons. Thus, 18 of the last 21 polygons are of measure 0, and the last 3 polygons are of negative measure. Consequently there is no distinction between the first 18 of these, and perhaps it is as well not to take too seriously the indication that the last 3 polygons are definitely worse than the others.

In the third place the formula is not a highly sensitive one, since the change produced by any single element of order is considerable. For example, all triangles whatsoever have an aesthetic measure of $7/6$, $2/3$, 0, $-1/3$, $-2/3$, or -1 according to the definition. No other gradations are possible.

If then, after laying aside connotations as far as possible, there is felt to be a gradual diminution in aesthetic value as the polygons are looked at in succession, so that polygons whose aesthetic measures are substantially unequal are properly arranged, while those of almost equal measures are of nearly equal attractiveness, the formula must be regarded as justified. We may conjecture in this event that the general theory is correct in its essential features. Of course considerable variations in individual judgments are to be expected.

In classes at Columbia University (summer, 1929) and Harvard University (summer, 1930) I obtained the consensus of aesthetic judgment as to the arrangement of these polygons. The results so obtained were found to be in substantial agreement with the arrangement obtained by the formula.

29. Possible Modifications

It is interesting to ask whether the appreciation of polygonal form has undergone a process of gradual evolution in the past, and whether further modifications are to be looked for.

The appreciation of symmetric form — as manifested by the human figure, by the sun, moon, etc. — and of equilibrium must have been present from the earliest times. The utilization of rectangular networks must go back at least as far as primitive architectural design. Hence it seems likely that the enjoyment of polygonal forms has not changed very much in character since the dawn of civilization.

If, for any reason, polygonal forms were to receive much attention, it seems certain that aesthetic appreciation of them would undergo further interesting development. We shall merely point to two obvious possibilities. From a mathematical point of view other uniform networks, such as that of equilateral triangles, have just as much geometric interest as the rectangular ones. As soon as the 'normal observer' becomes familiar with polygons closely related to networks of these new types, he will find that these polygons gain in aesthetic value. Polygons such as Nos. 51, 69, and 89 are of this kind. If such a development were to take place, the definition of the element HV of relation to a uniform network would require modification. Moreover, if polygonal forms came into greater use, more elaborate forms would become attractive. For example, perspectively correct representations of polygons found in the above list would be appreciated as such. This too would necessitate appropriate further modification in the definition of aesthetic measure.

30. The Mathematical Treatment of Aesthetic Questions

A complete theory such as that which precedes can be used as a logical tool in order to answer aesthetic questions by purely mathematical (logical) reasoning. For example, let us propose the following question: Which is the most beautiful of all polygonal forms?

POLYGONAL FORMS

To answer this question we observe first that the positive elements V, E, R, and HV cannot exceed 1, 1, 3, and 2 respectively in numerical value, while the most advantageous value of F is 0. In consequence O can never exceed 7, whence it appears that the aesthetic measure M of any polygon of complexity C cannot exceed $7/C$.

It is known that, for the square with one side horizontal, O is 6 and C is 4, so that M is 1.50. For any polygon for which C is as great as 5, it is clear that M will not exceed $7/5 = 1.40$. Hence we can conclude that any conceivable polygon for which M is as great as for the square has a complexity of 3 or 4.

But if the complexity C is 3, all the sides of the polygon lie on three straight lines. The only polygons of this kind are the triangles. However, for the scalene and isosceles triangles R and HV are 0, so that O is at most 2, and M is at most $2/3 = .67$. Hence these triangles can be excluded from consideration. For the equilateral triangle M is immediately found to be only $7/6 = 1.16\ldots$, so that it can be excluded as well.

It only remains to consider the possibility that the complexity C is 4, in which case the polygon lies wholly upon four straight lines. Such a quadrilateral cannot have network value ($HV = 2$ or 1) unless these four lines are two pairs of parallel lines; in this event it must be a square ($M = 1.50$), rectangle ($M = 1.25$), or diamond ($M = 1.00$). In any other case HV and either V or R are 0, and it is clear that O is at most 2 and M at most $2/4 = .50$.

It follows then as a 'theorem' that the square with horizontal sides with $M = 1.50$ is the best of all possible polygonal forms. Obviously such mathematical treatment upon the basis of the theory becomes a mere game if carried too far. It is only desirable to refer to this possibility of the theory in order to indicate its completeness.

31. On Uncertainty and Optical Illusions

In what precedes it has been assumed that the observer identifies the polygon completely from a mathematical point of view. Any consideration of the degree to which this assumption is actually valid would introduce difficult psychological questions.

Sometimes the degree of uncertainty is increased by certain systematic optical illusions. It is a well known fact, for instance, that there is a definite tendency to overestimate vertical distances as compared to horizontal distances. Because of this illusion, a rectangle not a square may be adjudged to be a square. In all of what follows, illusions and uncertainties, whether visual or auditory, will not be taken into account.

CHAPTER III

ORNAMENTS AND TILINGS

1. Ornaments and Motions

ANY figure, traced, painted, embossed, or otherwise executed on a surface, will be termed an 'ornament,' provided there is at least one possible motion which moves the figure, as a whole but not point for point, back to its initial position. We shall admit as legitimate 'motions': (1) the rotations about an axis through any angle; (2) the translations from one position to another having the same orientation in space; (3) the combination of motions of types (1) and (2) in a screw motion obtained by following a rotation about an axis with a translation along the axis; and also (4) the transference of each point to its mirror image in a plane, called reflection in a plane.

Evidently any polygon having symmetry furnishes a simple illustration of an ornament with one or more rotations (1). A uniform rectangular network is an ornament with infinitely many translations (2). A spiral staircase, thought of as indefinitely extended, forms an ornament (according to this definition) with infinitely many screw motions (3). A building which is constructed symmetrically on two sides of a vertical plane will also be an ornament in the technical sense, with a reflection in a plane (4).

From a higher mathematical point of view it is legitimate to consider such reflections (4) as motions in the following sense: if our space were immersed in a 'four-dimensional space,' it would be possible to move a figure into its mirror image, point for point, without deformation. A fully analogous case of a simpler type is the following: a general figure in a two-dimensional plane cannot be moved within the plane into its mirror image with reference to some line of the plane; but if the figure be rotated through 180° about this line in three-dimensional space, it will clearly be carried into its mirror image.

AESTHETIC MEASURE

Ornaments of this exceedingly general type are everywhere present in nature and in art.* Any figure which involves a homogeneous repetitive principle will usually be an ornament in the sense stated. This principle may be regarded as the outward, visible manifestation of the existence of such underlying motions, whether the ornament be a maple leaf or a spiral staircase.

For the sake of definiteness we shall limit attention to ornaments in a plane formed by a tracery in a single color or by a tiling. These form a typical and very important class of ornaments. The types of motions (1)–(4) admitted by such ornaments are the following: (1) rotations in the plane about a point; (2) translations in the plane; (3) slide-reflections obtained by a translation in the plane along a fixed axis and a half revolution about that axis; (4) reflection in an axis of symmetry (also produced by a half revolution about that axis).

2. Quasi-Ornaments

There are certain simple spiraliform figures which are not ornaments in the above technical sense, and yet produce an ornamental effect since

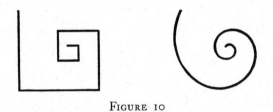

FIGURE 10

they appear to arise from a uniform repetitive principle. Such figures, which are illustrated herewith, will be termed 'quasi-ornaments.'

More precisely, let us define a 'quasi-ornament' as a figure formed by a curve or broken line which turns always to the left or to the right without touching or crossing itself, and in which the total rotation is at least through one complete revolution. It is interesting to inquire into the reason for the undoubted aesthetic effectiveness of such figures.

Just as a circle is the curve which embodies the suggestion of rotation, so the logarithmic or equiangular spiral (see Figure 10) is the curve which

---

* For a standard descriptive treatment of such ornaments, see J. Bourgoin, *Théorie de l'Ornement*, Paris (1883).

ORNAMENTS AND TILINGS

embodies the suggestion of proportionate diminution or enlargement. In fact if we diminish or enlarge this special curve in any selected ratio, it is seen to be the same curve *in size as well as in shape* as before. It was because of this strong suggestion of similarity that the mathematician James Bernoulli wished this curve inscribed on his tomb with the words "eadem mutata resurgo." Moreover, the suggestion of similarity induced by the equiangular spiral is carried over to other spirals, whether rectilinear or curvilinear.

Now in a mathematical sense which cannot be elaborated here, change in size of a figure can be regarded as analogous in certain fundamental respects to a motion of the figure. The following partial analogy at least is obvious: by a motion the given figure is placed in a new aspect; and by change in size much the same thing is brought about, since similar figures are ordinarily thought of as the same figure at a greater or less distance.

On this account, our theory of ornaments will be regarded as applying to these quasi-ornaments, and the presence of a quasi-ornament as part of a complete ornament will be considered as inducing the same element of order S which we shall attach to similarity of figures. The element S is the single essentially new positive element involved in our theory of ornaments, beyond the elements V, E, R, HV involved in polygonal form.

3. THE GROUP OF MOTIONS OF AN ORNAMENT

In dealing with ornaments, we shall regard the 'motion,' I, which leaves every point fixed, as a motion in the technical sense. By definition, there is at least one motion other than I which carries an ornament into itself. If A denotes any one of this set of motions, clearly the motion of repetition of A, which may be denoted by AA, also moves the ornament into its first position. Furthermore, if B denotes any other motion whatever, the combination AB of the motion A followed by B is also such a motion. The totality of motions carrying the ornament into itself will be called its 'group of motions' (cf. Chapter II, section 23).

If F be any part of an ornament which takes the positions $F, F', F'',$ ... under the various motions of the group, then $F, F', F'', \ldots$ will be said to be 'of the same type'; also if a region F together with the other regions of the same type fills the plane without overlapping, then F will be said to be a 'fundamental region.'

The portion of the ornament lying in any fundamental region suffices to generate the entire ornament by its repetition under the motions of the group, and may be called a 'fundamental portion' of the ornament. This is of course to be distinguished from the 'repeat' of the ornament, out of which the ornament arises by repetition from the various translations of the group.

Conversely, any figure whatever lying in a fundamental region of any group will yield an ornament when reproduced in all the positions of the same type assigned to it by the given group. It may happen, however, that the ornament so obtained admits of motions into itself besides those in the given group, and so possesses a still larger group.

4. Classification of Ornaments

An obvious basis of classification of plane ornaments is as rectilinear, curvilinear, or mixtilinear, according as they are made up of straight lines, curved lines, or a combination of straight and curved lines. The quasi-ornaments in Figure 10 are rectilinear and curvilinear respectively; the semicircle furnishes a simple instance of an ornament of the mixtilinear type.

A more fundamental method of classification is furnished by means of the underlying group. In fact ornaments may be divided into two main classes: those for which the underlying group contains only a finite number of distinct motions and so is called 'finite,' and those for which the group is 'infinite.'

All ornaments with a finite group will be called 'simple' ornaments. Symmetric polygonal forms illustrate this case. The circle, as well as sets of concentric circles, will also be regarded as simple ornaments, in spite of the fact that the group of motions contains the infinitude of rotations about the center and of reflections about any diameter.

The remaining ornaments, having an infinite group, can be subdivided further into two mutually exclusive classes: the 'one-dimensional' or band ornaments, in which a fundamental figure is repeated in a one-dimensional array so that some line — the 'axis of the ornament' —

ORNAMENTS AND TILINGS

remains fixed in position under any motion of the group, and the 'two-dimensional' or all-over ornaments, in which a fundamental figure is repeated in a two-dimensional array, and no line is fixed in position under all the motions of the group.*

Only those ornaments will be considered which are made up of straight lines and ordinary curves; the one- and two-dimensional ornaments formed by a set of parallel straight lines will be left out of further consideration as of no especial aesthetic interest.

The uniformly spaced o's:

$$\ldots \text{o o o o o o} \ldots$$

constitute a one-dimensional ornament with axis passing through the central points. Similarly the plane array of T's:

$$
\begin{array}{c}
\cdot\ \cdot\ \cdot\ \cdot\ \cdot\ \cdot \\
\cdot\ \cdot\ \text{T T T T}\ \cdot\ \cdot \\
\cdot\ \cdot\ \text{T T T T}\ \cdot\ \cdot \\
\cdot\ \cdot\ \text{T T T T}\ \cdot\ \cdot \\
\cdot\ \cdot\ \cdot\ \cdot\ \cdot\ \cdot
\end{array}
$$

constitutes a two-dimensional ornament. It is worth while to consider these examples further for illustrative purposes.

In the one-dimensional example the corresponding group of motions is constituted by the following: translations to right or left by any multiple of the distance between the centers of two adjoining o's; reflections in the axis itself; reflections about any line perpendicular to the axis and passing through the center of some o, or passing midway between two centers; the slide-reflections which arise from combination of the first two types of motion, or which may be thought of as produced by a single screw motion with angle of rotation 180°.

For the two-dimensional ornament the group of motions consists of the following: translations to right or left by the distance between two columns of T's; translations up or down by the distance between two rows of T's; translations which are formed by combinations of these two types; reflections about the vertical axis of symmetry of a column of T's, or

---

* For the mathematical basis of this and subsequent assertions, the reader is referred to A. Speiser, *Theorie der Gruppen von endlicher Ordnung*, second edition, Berlin (1927), in particular chapter 6.

midway between two columns; slide-reflections about these same axes in which the slide is up or down by any multiple of the distance between two successive rows.

In fact, to restore the two-dimensional ornament to its original position as a whole, any letter T must be made to coincide with some other T either with the same orientation or reversed. In the first case the motion involved is clearly a translation. In the second case, if the second letter coincides with the first, a reflection about the vertical axis through the letter is required. If the second letter is another letter in the same column of T's, a slide reflection about the axis of that column is required. If the T is transferred to some other column, either a reflection or slide-reflection about the vertical axis midway between the two columns is evidently required.

The notions of figures 'of the same type,' of 'fundamental regions,' and of 'fundamental portions' of the ornament can also be illustrated by these examples. Thus in the one-dimensional ornament the letter o's are figures of the same type; so are also the set of upper and lower halves of these letters, the set of right and left halves, and the set of quarters into which these letters are cut by their horizontal and vertical axes of symmetry. Similarly all the T's are of the same type in the second ornament; so are also the set of right and left halves of these letters.

A simple fundamental region in the first case consists of a part of the plane above and to the right of the center of some o but to the left of the mid-point between it and the following o; a fundamental portion of this ornament is formed by a quarter of any letter o. In the second case a simple fundamental region consists of a rectangle occupied by the part of a line of type between successive vertical axes of symmetry; a fundamental portion of the ornament is formed by half of any letter T.

A further classification of one- and two-dimensional ornaments into species may be made according to their groups of motions. Two groups will be said to be of the same 'species' in case they can be defined by geometrically similar ornaments undergoing similar motions, or at any rate by ornaments similar except for a proportionate reduction of distances in a single direction.

ORNAMENTS AND TILINGS

For example, the groups attached to any two uniform diamond (not square) networks are of the same species, since one diamond network can be deformed to be similar to the other by such a reduction. On the other hand the group of a square network is not of the same species as that of any diamond network. In fact the first may be rotated through 90° into itself about the center of any square, while there is no similar motion in the case of the diamond network.

5. On the Determination of Species

The species of the simple ornaments and their finite groups have been given in the preceding chapter (section 23). There are precisely 7 species of one-dimensional infinite groups and 17 species of two-dimensional infinite groups, all illustrated by appropriate rectilinear and curvilinear ornaments in the Plates VIII–XII, opposite the next page. The proof that this classification into species is exhaustive is a matter for careful mathematical discussion which cannot be entered upon here.*

A practical question arises, however, as to how the species of a given ornament is to be positively identified as being one of the listed species. A little experience makes the practical determination of species an almost intuitive affair; but for the theoretic determination it is necessary to search for that particular one of the listed ornaments which admits essentially the same types of motion as the given ornament, and no others.

6. Ornaments and Ornamental Patterns

An 'ornamental pattern' may be described as an ornament containing a collection of ornaments arranged in a spatial hierarchy, but so ornate that the whole of the pattern cannot be grasped at a glance. Such a pattern may be found to be exceedingly interesting as the eye passes from the principal to the secondary ornaments, appreciates them in turn, and notes their interrelations. The favorable impression thus obtained results from a summation of the aesthetic effects of the constituent ornaments.

* Cf. Speiser, *loc. cit.*, who follows the classification of Polyá adopted here. It may be noted that in the illustrations, the rectilinear and curvilinear two-dimensional ornaments of all species are in the same orientation, except for species XIII, XVI, and XVII, in which horizontal and vertical are interchanged. In order to secure an example of a rectilinear ornament of species IV, an Egyptian mat design has been slightly modified.

AESTHETIC MEASURE

Now while the elaborate ornamental pattern may be a consummate work of art, it is not to be regarded as a good ornament. On the contrary it must be regarded as an unsuccessful one, because, relatively to the elements of order, O, *appreciated at a glance*, the complexity, C, is very large. The mosaic shown earlier (Plate I) furnishes an illustration of an ornamental pattern.

7. The Aesthetic Problem of Ornaments

In our treatment of polygonal forms it was found to be desirable to fix attention upon polygonal tiles of uniform size, color, and material, and set in a favorable vertical position. By this device all connotative factors were practically eliminated.

An equally satisfactory treatment of plane ornaments cannot be anticipated. The types of such ornaments are so varied that it is not pos-

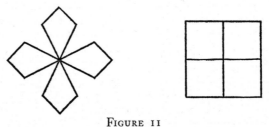

Figure 11

sible to find all of them used for similar purposes. Moreover such ornaments are in general superficial decorations (by definition), and are determined in character by their specific environment. In consequence they do not possess independent aesthetic value to the same extent as the more primitive polygonal forms. Finally, just because of these facts, they are likely to involve connotations. For example, the first of the two simple ornaments above will suggest a conventionalized flower, and the second a window grating.

Notwithstanding these difficulties, we shall attempt a theory of the aesthetic measure of simple rectilinear ornaments; but in the case of one- and two-dimensional ornaments, we shall only develop such a theory for the special two-dimensional ornaments formed by tilings, which evidently constitute the simplest possible kind. It would not be difficult to provide an analogous theory for the one-dimensional ornaments formed by tilings.

PLATE VIII

The 7 Species of One-Dimensional Ornaments

PLATE IX

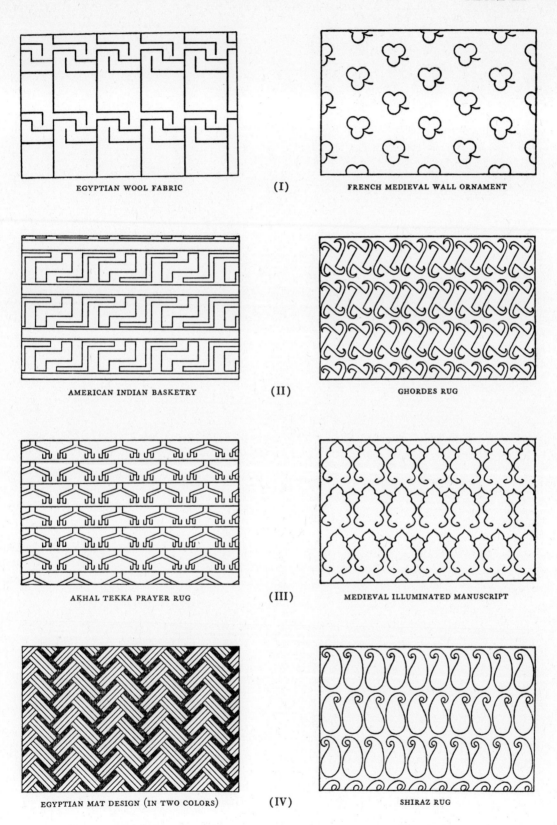

THE 17 SPECIES OF TWO-DIMENSIONAL ORNAMENTS, NOS. I–IV

PLATE X

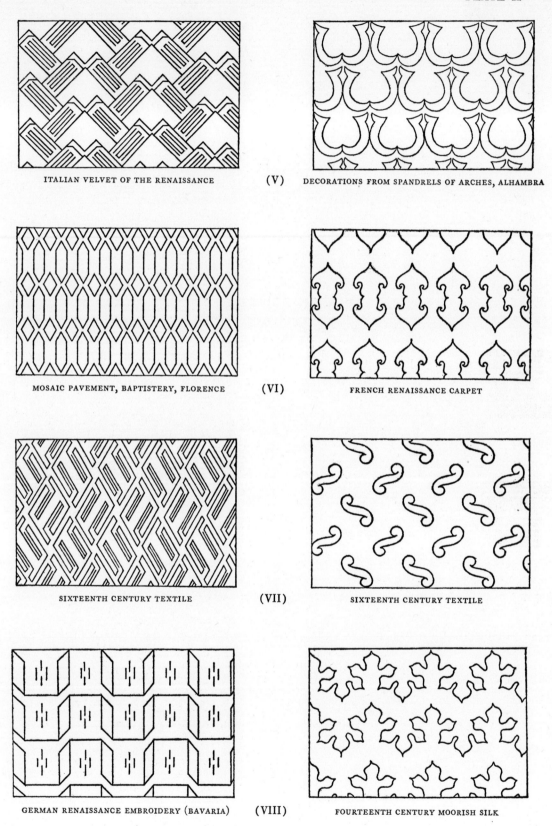

THE 17 SPECIES OF TWO-DIMENSIONAL ORNAMENTS, NOS. V–VIII

PLATE XI

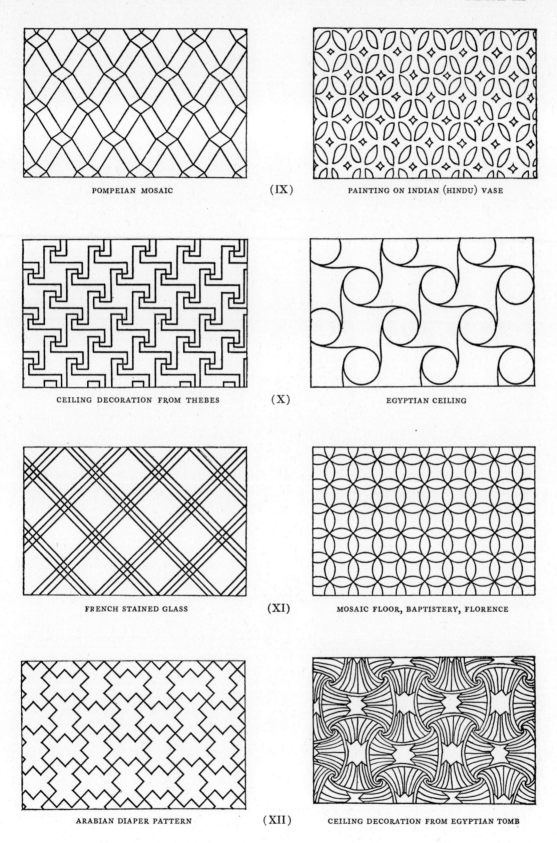

THE 17 SPECIES OF TWO-DIMENSIONAL ORNAMENTS, NOS. IX–XII

PLATE XII

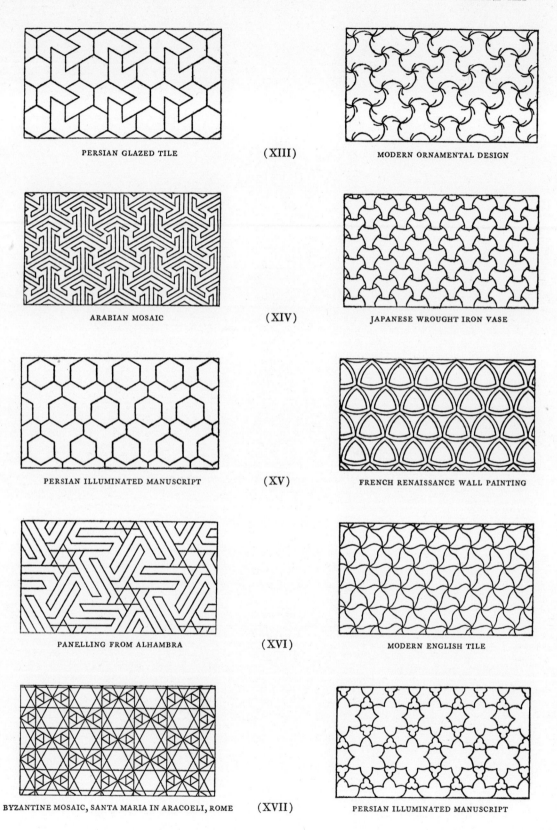

THE 17 SPECIES OF TWO-DIMENSIONAL ORNAMENTS, NOS. XIII–XVII

ORNAMENTS AND TILINGS

8. Polygons as Simple Rectilinear Ornaments

A theory of the simple rectilinear ornaments formed by polygons has been elaborated in Chapter II. It is true, of course, that the precise problem treated there was that of polygonal form realized in similarly colored tiles of uniform size and material, set in a vertical plane, while the polygonal ornament is usually thought of as merely traced in outline in such a plane. However, the aesthetic factors involved are essentially the same, whether we consider a polygonal tile or its outline traced in a plane. There is only a slight difference in so far as the element of equilibrium, E, is concerned; for, a tracery is less easily felt to be out of equilibrium than a massive tile.

We shall overlook this slight difference and consider that polygonal ornaments have an aesthetic measure which may be determined by the same rules as those adopted for polygonal tiles. Thus our list of 90 polygons furnishes us also with a classification of the corresponding simple polygonal ornaments in order of decreasing aesthetic measure. It is true that some of the best simple rectilinear ornaments are given by these polygonal forms. Nevertheless, other types are decidedly attractive in many cases, and are readily compared intuitively in regard to their aesthetic quality. In the section which follows we shall develop a modification of the theory of polygonal form, applicable to any simple rectilinear ornament.

9. Aesthetic Measure of Simple Rectilinear Ornaments

Obviously such a theory of the simple rectilinear ornament must involve all of the factors observed in polygonal form. In our theory there appears only a single new element of order, namely the element S of similarity already referred to. If this theory is found to agree fairly well with the aesthetic judgment, to the extent that such judgment is definite, it will have accomplished all that can be reasonably expected.

As we have just indicated, the basic formula for the aesthetic measure M of the simple rectilinear ornament will be expressed in the form:

$$M = \frac{O}{C} = \frac{V + E + R + HV + S - F}{C}$$

where V, E, R, HV, F, and C are natural modifications of the corresponding elements in the case of polygons, and where S denotes the single new element of similarity.

C

The complexity C is defined precisely as in the case of polygons, namely as the number of straight lines containing at least one side of the ornament.

V

The element V of vertical symmetry is also defined precisely as before, namely as 1 or 0 according as the ornament is or is not symmetric about a vertical axis.

E

We shall consider the ornament to be made of thin uniform wire in determining its center of area. If we regard the separate parts of the complete ornament as held together rigidly, we may obviously define the horizontal line segment AB supporting the ornament from below just as in the case of a polygon. Thus we may define E to be 1, 0, or -1 as before.

R

The element R of rotational symmetry is defined as before, excepting that only polygonal niches of the ornament are considered.

HV

The definition of the element HV will not be modified, excepting that when there are at least two polygons in the ornament, two exceptions instead of one of either type (Chapter II, section 27) will be allowed in the case $HV = 1$.

S

By a 'polygon of the ornament' we shall mean (1) the 'elementary polygons' bounded by sides of the ornament, not touching any part of the ornament within them; (2) the outermost polygons of the ornament, accessible from the outer part of the plane; and also (3) the polygons made up of sides of the ornament, whose vertices are not the end points of further sides of the ornament.

It is these polygons which are noted by the eye. For example in the hexagram (listed as No. 1 in Plate XIII, opposite page 60) the central regular hexagon and the six small equilateral triangles which abut on its sides are elementary polygons of the first type; the outermost six-pointed star is of the second type; and the two large equilateral triangles are of the third type. In this case there are ten polygons of the ornament.

The new element S of similarity will be defined to be 0 if no 'polygon of the ornament' and polygonal figure not of the same type (section 3) are similar and in the same orientation, and if no quasi-ornament (section 2) is present.

If there is only one such relationship of similarity, either between two polygons not of the same type, or due to a single type of quasi-ornament, S is defined to be 1; and if there is more than one such relationship, S is defined to be 2. These types of similarity are obviously pleasing to the eye. The reason why the presence of a quasi-ornament induces an associative reference to similarity has already been explained.

F

This element of unsatisfactory form will be made up of a part F_1 defined just as for the case of a polygon, and another part F_2 dependent on a number of new aesthetic factors which are defined below.

In the determination of F_1 the term 'unsupported re-entrant side' will only be used for a side of the ornament which is an unsupported re-entrant side (in the previous sense) of every elementary polygon of which it is a side, which cuts at least one elementary polygon when extended, and which does not extend into any entirely distinct side of the ornament. A 'niche' of the ornament will be a niche for some elementary polygon or for some complete part of the ornament.

The second part F_2 of F is designed to include the essentially new kinds of negative elements of order which are found in simple rectilinear ornaments other than polygons. These will be denoted by $F_2(a)$, $F_2(b)$, ... $F_2(i)$. Only in case none of these are found in the ornament will F_2 be 0; furthermore F_2 will never be taken to exceed 4. These elements are defined respectively as follows:

(a): index 1. (a) is counted once for each type of line containing at least one side which abuts perpendicularly, without continuation, on some side of a polygon of the ornament.

(b): index 1. (b) is counted once for each type of vertex at which terminate an odd number of sides, and once for each type of free end other than those which terminate a quasi-ornament.

(c): index 1. (c) is counted (only once) if the center of area of the ornament (considered as made of uniform wire) lies on a side of the ornament and is within a polygon of the ornament.

(d): index 2. (d) is counted once for each type of line occupied by an interior side of the ornament, which is not parallel to any side of the outermost polygon which encloses the side nor to any side forming part of the minimum convex polygon enclosing the entire ornament.

(e): index 2. (e) is counted once for each type of elementary polygon containing within it a part or parts of the ornament whose group of motions (section 3) does not contain all the motions in the group of the part of the ornament to which the elementary polygon belongs.

(f): index 2. (f) is counted once for each type of polygon of the ornament containing within it only similar elementary polygons in a different orientation.

(g): index 2. (g) is counted as the number of types of parts of the ornament entirely outside of each other, in case such exist.

(h): index 2. (h) is counted once for each type of elementary polygon of the ornament which is attached from the outside to the rest of the ornament only where the vertices of the polygon meet the interior points of sides of the rest of the ornament.

(i): index 3. (i) is counted if the minimum enclosing polygon about the entire ornament meets the ornament only at free end points.

10. Some Explanatory Comments

The rôle of the element S of geometrical similarity has already been explained.

In the 30 simple rectilinear ornaments of Plates XIII and XIV opposite, the element S enters in Nos. 1, 3, 4, 7, 8, 9, 10, 11, 12, 17, 18, 22, and 23, with S equal to 1 or 2 as the case may be. Of these ornaments, Nos. 1, 2, 3, 4, 5, 9, 10, 11, 12, 13, and 15 have satisfactory form (that is, F is 0).

The same list of ornaments gives examples of the element F_1 in Nos. 17, 18, 21, 23 in which there are two types of niches, No. 19 in which

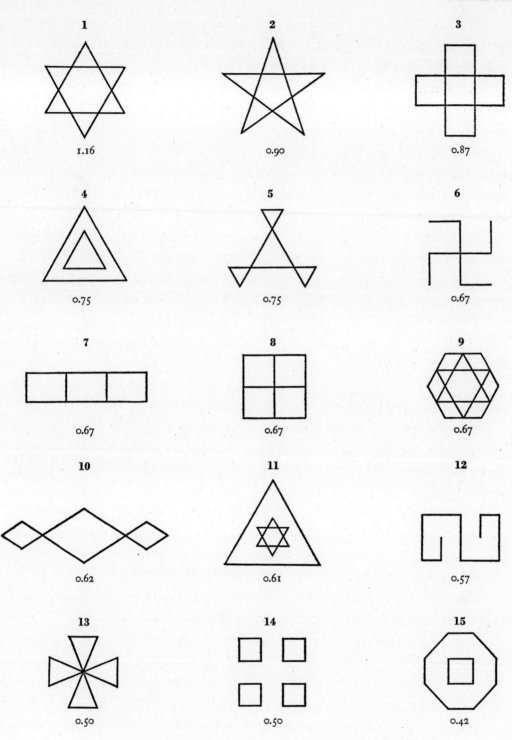

AESTHETIC MEASURES OF 30 SIMPLE ORNAMENTS, NOS. 1–15

PLATE XIV

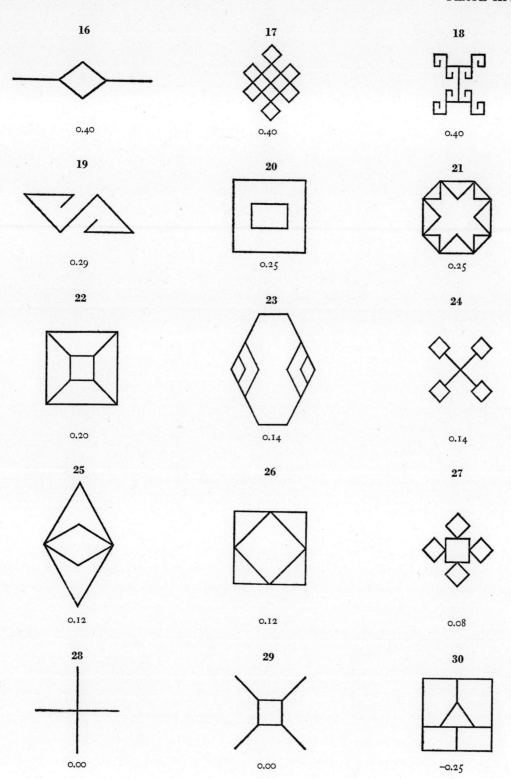

AESTHETIC MEASURES OF 30 SIMPLE ORNAMENTS, NOS. 16–30

ORNAMENTS AND TILINGS

there are three types of directions, and No. 24 in which there is one type of unsupported re-entrant side.

The reasons for the various new negative elements of order found in F_2 may be summarized as follows:

ad (a). When a side abuts perpendicularly upon an interior point of a line of the ornament, there is a feeling of dissatisfaction, for the eye in traversing the side is doubtful as to which direction to follow. This effect disappears, however, if the end point is not part of a polygon, since then the desire to follow the ornament systematically with the eye seems to disappear. By comparison of No. 18, for which $F_2(a)$ is 0, with Nos. 7, 8, 24, for which $F_2(a)$ is 1, and with No. 30, for which $F_2(a)$ is 2, this assertion is seen to be reasonable.

ad (b). The presence of vertices at which an odd number of sides terminate is distinctly disadvantageous, because the eye is held to such a vertex; with an even number of sides the eye can follow the sides through the vertex in pairs, and in this manner take account of them all without stopping at the vertex. The same difficulty is felt with free ends not terminating a quasi-ornament, as with other odd vertices.

Ornaments Nos. 6, 7, 8, 16, 18, 22, 23, 24, 28, 29, and 30 illustrate the case when $F_2(b)$ enters.

ad (c). A tracery suggests a line rather than a filled surface. Hence the center of area is defined as stated.

In case this center of area lies within a polygon of the ornament, the eye looks for a center of repose at this point; but if sides of the ornament pass through the point, the eye seems to be drawn away in these directions to the enclosing polygon, so that the result is not satisfactory.

On the other hand, if the center of area lies outside of all the polygons of the ornament, the eye seeks a center on a line of the ornament, so that the presence of such lines is an advantage rather than otherwise. The only ornaments of the list for which $F_2(c)$ is not 0 are Nos. 8 and 17.

ad (d). An interior side of the ornament of the type described in (d) is felt not to be properly suited to the rest of the ornament, because its direction is not strongly suggested by the leading directions of the ornament. Nos. 22, 25, 26, 27, and 30 illustrate the effect of incongruity thereby produced.

ad (*e*). In case the interior part of the ornament referred to in (*e*) admits a smaller group of motions than the enclosing ornament, this part is felt not to be suitable. For this reason an ornament formed by a rectangle within a square, as in No. 20, is much less satisfactory than one formed by a square within a rectangle. No. 20 is the only ornament of the list for which $F_2(e)$ is not 0.

ad (*f*). When, for instance, an equilateral triangle is placed within another such triangle, with reversed orientation, there will be observed the effect taken account of in $F_2(f)$. This element $F_2(f)$ occurs only in No. 26 of the list.

ad (*g*). The element $F_2(g)$ is 0 unless the ornament has entirely distinct parts outside of one another. This produces a lack of unity in the ornament, illustrated in a simple way by No. 14.

ad (*h*). The element $F_2(h)$ seems to have a utilitarian origin. If a polygon is attached to the rest of the ornament in the manner described, the attachment is felt to be insecure. This possibility is illustrated in No. 26 only.

ad (*i*). In case $F_2(i)$ is not 0, there is a strong suggestion that the ornament would be difficult to handle, because of the sharp free ends which jut from it. The only ornaments of the list of this type are Nos. 28 and 29.

11. Application to 30 Simple Rectilinear Ornaments

The 90 polygonal forms listed in Chapter II have the same aesthetic measure whether considered as tiles or as ornaments traced on a surface. The 30 ornaments shown in Plates XIII and XIV are not of this simple type, but are chosen to illustrate the application of the extended definition of aesthetic measure given above to a fairly wide range of ornaments. The 30 ratings so obtained are to be thought of as intercalated among those of the 90 polygons rated earlier.

The reader should find a gradual diminution in aesthetic value, according to the measure M obtained, if the definitions made are justified. More precisely, it should be found that almost all pairs of ornaments selected from the list are arranged in proper order. Any effect of novelty in these forms must be discounted. Moreover, connotations must be taken into consideration. In particular the ornament No. 8 is likely to suggest a window grating, and this connotation is not an agreeable one.

The ornament No. 15 would not be nearly as satisfactory if the outer octagon were regular, and its aesthetic measure would then be only .16. In fact by this modification HV would be reduced from 1 to 0, and F would increase from 0 to 2 on account of the presence of the element $F_2(e)$ of index 2. Because of the fact that the actual octagon of No. 15 is almost regular in the given ornament, there is thus a slight unfavorable effect of ambiguity in this case, which is to be explicitly borne in mind.

The ratings assigned in these cases, and in others which I have tried, seem to be approximately correct. The reader who wishes to check the theory should apply it further to a diversified set of ornaments, and so supplement the results tabulated. Evidently the negative elements of unsatisfactory form in F_2 form a somewhat miscellaneous collection. It is entirely possible that other similar elements have been overlooked.

12. Other Types of Ornaments

The only curves which can be regarded as fully identifiable would appear to be the circle, the ellipse which may be interpreted as a circle seen in perspective, and the parabola, which is a limiting elliptic form. These geometrical curves, like the straight line, are determined by a certain number of their points: the circle, by the two extremities of any diameter; the ellipse, by the four extremities of its diameters; the parabola, by its vertex and focus. Other curves are not wholly determined from the intuitive point of view by any number of their points, no matter how large. There are comparatively few curvilinear and mixtilinear ornaments which are made up of these special geometrical curves.

In consequence, most curvilinear ornaments ordinarily used must be regarded as of infinite complexity from a geometrical point of view, since they are not determined by any finite set of points. This general deficiency is remedied in practice by the use of parts which represent some selected object of pleasing type in a simple conventionalized manner; the objects employed are mostly floral in character. In virtue of this symbolic quality, the curves of the ornament acquire an identifiable form which

AESTHETIC MEASURE

would otherwise be lacking, and this circumstance operates to counteract the feeling of indeterminateness.

It will be understood, then, why no theory of the aesthetic measure of curvilinear or mixtilinear ornaments is practicable, for, besides the purely geometrical factors analogous to those involved in the simple rectilinear ornaments, there are nearly always connotative factors of importance.

Moreover, in the case of rectilinear one- and two-dimensional ornaments, the effect produced is usually that of an ornamental pattern (section 6) rather than that of an ornament. Here the eye takes in separately the various constituent simple ornaments of the pattern, and the total effect produced is felt as a kind of average.

In this case also, the number of constituent ornaments is usually so large, and their effectiveness is so conditioned by the suitability of the ornament to its environment, that a theory of aesthetic measure hardly appears to be practicable.

13. The Aesthetic Measure of Tilings

Any two-dimensional rectilinear ornament made up of polygonal forms may be regarded either as a tiling or as a pattern of tiles set in stucco. In Plates IX–XII, 11 of the 17 species, namely IV, VI, IX–XVII, fall in one or the other of these categories. Now in ordinary tilings, such as are illustrated also in the opposite Plate XV, intuitive comparisons of aesthetic value are found to be possible.

It may be anticipated then that the aesthetic measure of tilings * may be defined in accordance with our general theory. In attempting to do this we shall assume for definiteness that the tiling is set in a vertical plane. We assume also that it is symmetric about the vertical direction (that is, is the same as its mirror image). This requirement is almost always met in practice, as for instance in the illustrations opposite.

In setting up a theory which agrees with the observed facts, it is found necessary, however, to regard the complexity C as determined by the number of tilings in a certain 'fundamental visual region' F, and the order O as determined by the aggregate of the aesthetic measures of certain 'interesting polygons' of this region. The psychological explanation lies

* Or, equally well, of leaded glass windows.

PLATE XV

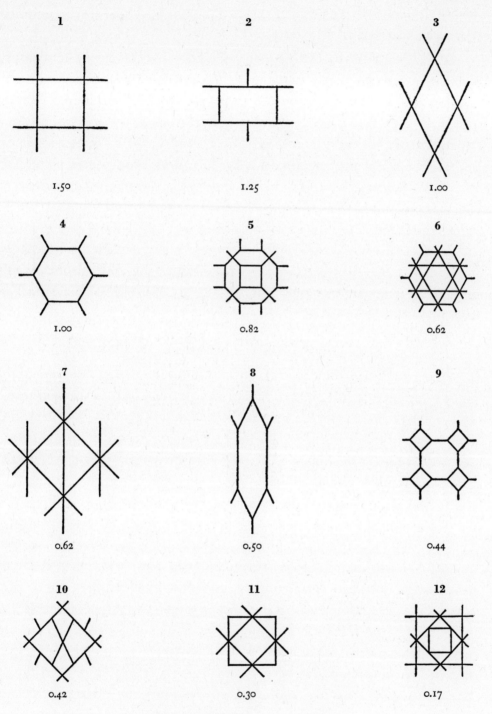

AESTHETIC MEASURES OF 12 TILINGS

apparently in the fact that the tile is felt as an ornamental pattern rather than as a single ornament, whose success is measured by the average aesthetic quality of the constituent polygons. These polygons appear to be rapidly identified *seriatim*, but not to be enjoyed thoroughly as isolated polygonal forms. Let us proceed to the detailed formulation of such a theory.

In looking at the tiling we can find a 'fundamental visual region' F, made up of a least number of the polygonal tiles, and possessed of the two following properties: (1) F admits as nearly as possible the types of symmetry found in the tiling; (2) the region F and the regions F', F'', ... obtained from it by the translations of the group of motions cover the entire tiling.

In order to define the order O we specify the 'interesting polygons' to be found in the region F. These are of two kinds: the elementary polygons (not triangles) in F possessing rotational symmetry and equilibrium; the 'central polygons' in F which admit the group of motions of F (and perhaps others). Among these latter will always be counted the polygon formed by the outline of F.

For example, in the tiling No. 5 of those seen in Plate XV, the square tile and F itself are the central polygons. In this case all of the elementary tiles have the requisite qualities. On the other hand, in the tiling No. 6 the hexagonal tile, the six-pointed star, and the hexagonal outline of F are the central polygons; besides these, only the remaining two diamond polygons with vertical symmetry enter into consideration. It will be observed that the sharp triangular forms are not of interest in such a tiling.

We define the order O as the sum of the aesthetic measures of the 'interesting polygons,' in which rectangles and squares are counted once only (no matter how often they appear in F), and other polygonal forms are counted once or twice in each orientation according as they appear once or are repeated. Mere differences in the sizes of these polygons are disregarded. The reason for these limitations is that the square and rectangular shapes seem to add no new interest through repetition in F, while other forms may be repeated at least once to advantage.

The aesthetic measure M is of course defined as the ratio O/C.

14. Application to 12 Tilings

In the preceding Plate XV are shown 12 tilings, arranged in order of decreasing aesthetic measure M according to the theory just outlined. The fundamental visual region F is shown in each case. In Nos. 1, 2, 3, 4, and 6, where F consists of a single polygon, the aesthetic measure is simply the measure of this polygon according to the theory of Chapter II.

It will be observed that all of these selected tilings have symmetry in other directions than the vertical, and hence admit of other possible orientations. The best orientations have been chosen in all cases where there is a difference in the aesthetic measures. However, in the other cases, where the orientation is theoretically indifferent, there will be found to be some preference for the orientation which stresses the vertical direction; for instance such is the case in the diamond tiling, No. 3. This is doubtless because these same ornamental forms are often employed in the leaded glass windows of churches, where there is a general desire to emphasize the vertical direction.

CHAPTER IV

VASES

1. The Problem of the Vase

IT is well known that skilful potters as well as connoisseurs attach high importance to the form of a vase. Various other aesthetic factors may enter, such as archaeological significance, decoration, and surface quality. But, more than anything else, it is their superior geometric form which is felt to characterize the best vases.

The definitive aesthetic quality of the form of many Greek vases has led Pottier[*] to say, "The proportions of vases, the relations of size between the different parts of pottery, seem to have been the object of minute and delicate researches among the Greeks. . . . If a profound study were made of it, perhaps one would find a system of mensuration analogous to that of statuary." His suggestion would ascribe the beauty of the vase to certain specific geometric proportions, such as Michelangelo conceived to be present in the ideal human body. This type of explanation derives a certain reasonableness from a vague analogy between the shape of the vase and of the human body — an analogy borne out by the usual descriptive terms, 'lip,' 'neck,' 'shoulder,' and 'foot,' applied to vases. Others would ascribe the mysterious beauty to the 'living form' of the graceful contour lines. Very recently a Pythagorean explanation, based upon supposed mystic virtues of irrational numbers, has been proposed by Hambidge.[†]

Now the systematic point of view which we have adopted does not permit us to take refuge in an obscure similarity between the shape of a vase and of the human body, nor in an occult quality of living form, nor even in mystical properties of numbers. Instead we are committed to a straightforward objective course in dealing with the problem of the vase,

---

[*] *Catalogue des vases antiques de terre cuite*, Paris (1906), vol. 3, pp. 658–659, my translation.
[†] *Dynamic Symmetry of the Greek Vase*, New Haven and New York (1920). See also L. D. Caskey, *Geometry of Greek Vases*, Boston (1922), for an important supplement to Hambidge's suggestive work.

namely to determine a suitable aesthetic measure, M, of the geometric form, based upon the relations of order, O, really appreciated by the eye, and a reasonable estimate of the complexity, C, of the vase.

Our attention will be directed exclusively to vases without handles or other asymmetric parts. The absence of such parts simplifies the aesthetic problem of the vase, since it is a pure geometric solid of revolution under these circumstances, whose form is completely determined by the shape of cross-section.

Vases are used more frequently for purely decorative purposes than other similar objects such as goblets, jars, etc. Nevertheless the theory advanced can be applied without essential modification to these aesthetic objects. In such cases the utilitarian requirements are changed of course, and must not be lost sight of.

2. The Method of Attack

The problem of vase form is a very difficult one, and the theory obtained here is to be regarded as more questionable than the theories of polygons and ornaments given in the two preceding chapters. In fact it is conceivable that, aside from certain obvious requirements such as regularity of curvilinear outline and general suitability, there is no genuine aesthetic problem of the vase. At any rate I should expect that *any* vase form meeting these elementary requirements would appeal to *some* intelligent person, if only by virtue of its novelty.

On the other hand, the definite appreciation shown by connoisseurs for particular vase forms seems to point to the probability that these do admit of intuitive aesthetic comparison.

According to our general theory, then, there must exist certain elements of order O which are appreciated by the eye, upon which depend the aesthetic appeal; and in terms of these elements of order O and the complexity C of the vase, it should be possible to define a suitable aesthetic measure M.

Evidently the primary question to be answered is: What are the appreciable elements of order which are found in the form of vases? It is peculiarly necessary that the reasons for the particular choice of these elements of order, among infinitely many other conceivable ones, be fully explained; for if these are not properly chosen, the whole theory is without basis. Once this question has been answered, we can proceed to deal with the secondary questions of regularity of form and suitability, and to define the aesthetic measure M.

3. The Symbol of the Vase

We shall adopt the plane figure formed by the *visual contour* of the vase as a symbolic representation of the vase. It is supposed of course that the vase is regarded from the most favorable position. This symbol has the merit of being what one actually sees. Moreover, at a reasonable distance away, this figure becomes practically that of a cross-section of the vase. Nevertheless the contour seen in perspective always differs slightly from the cross-section.

The contour will consist of the two curvilinear sides, symmetric with respect to the axis, and of two convex elliptical ends, one of which will become rectilinear when the observer is on a level with the lip or foot of the vase. These ends offer then no special interest as curves, since they are invariably the same.

Thus the theory will refer exclusively to the visual contour. In consequence, if it were desired to construct a vase with a given symbol, it would not suffice to give the vase precisely the cross-section of the symbol; instead it would be necessary to effect a correction so that the visual contour takes the prescribed form at a suitable distance.

In justification of this choice of the visual contour rather than of the contour of cross-section, it may be remarked that a coin seen in perspective is appreciated as an elliptical form and not as a circular one.

4. Characteristic Points and Tangents

There are certain points of the contour of the vase upon which the eye can rest and which play a vital part in the theory to be proposed. These are: (1) the points of the contour line where the tangent is vertical; (2) the points of inflection where the curvature changes direction from concave to convex; (3) the end points of the contour; (4) the corner points where the direction of the tangent changes abruptly. These four types of points will be called 'characteristic points,' and the corresponding tangents at these points will be called 'characteristic tangents.' The vase form of Figure 12 shows all four kinds of characteristic points, and the corresponding characteristic tangents.

Our claim that these particular points and directions play a vital aesthetic rôle is based upon the following considerations.

It was observed in the preceding chapter that curvilinear and mixtilinear ornaments, other than those which involve only the simplest geometrical figures, namely straight lines, circles, ellipses, and perhaps also parabolas, are infinitely complex by reason of the fact that no finite set of points determines the curves completely. In practice this deficiency is usually met by employing only such ornaments as involve conventional representation, mainly of foliage and flowers, or have other definite connotations. By this means the feeling of indeterminacy inherent in such ornaments is eliminated.

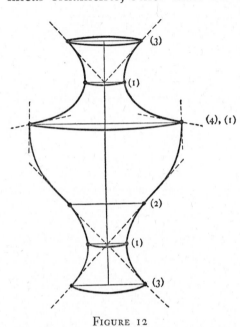

FIGURE 12

In the contour curves of vases, however, these simple geometric curves are not employed in general, although they are in rare instances; a goblet with parabolic contour is shown in Figure 13.* Furthermore there are in general no connotative elements in the contour curves of vases.

If then we concede that the problem of curvilinear form in vases is analogous to that involved in ornaments, we are forced to the conclusion that the simple curve of the contour must be regarded as practically determined by certain of its points, and that some precise geometric configuration must be suggested by the symbol of the vase, the relations of whose parts determine the elements of order.

It is apparent that the axis of the vase, represented by the axis of the symbol, is one fundamental part of this configuration. Furthermore, the

* I am indebted to Professor Speiser for the illustrative photograph of this Venetian goblet.

VASES

characteristic end points of the sides, the characteristic points of vertical tangency, and the other characteristic points of the contour form parts of this configuration. It does not seem, however, that the eye can identify any other points on the curve of contour. Moreover, the characteristic tangents at these points are determinate; in particular the characteristic points of vertical tangency are identified by the fact that the tangent is vertical.

If now we connect corresponding characteristic points by horizontal lines, and construct lines parallel to the axis through these points, we obtain the 'characteristic network' of the vase. This network, together with the characteristic tangents, will be considered as the specific geometric configuration attached to the symbol of the vase.

This choice is justified by the fact that the characteristic points and characteristic tangent lines are immediately suggested by the inspection of the visual contour, as has just been explained.

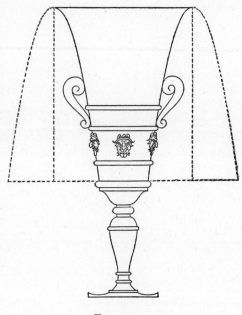

FIGURE 13

5. THE APPRECIABLE ELEMENTS OF ORDER

If two or more horizontal distances across the vase at these characteristic points are equal, the relationship tends to simplify the characteristic network and to unify the vase form. For example in the vase of Figure 12 the breadth of the vase is the same at the base, at the inflectional points, and at the top. In the same vase the breadth at the lower points of vertical tangency is the same as that at the neck.

Similarly a relationship of two to one in these horizontal distances is appreciated. Perhaps this is because the axis of the vase bisects all of these lines. In the same vase relations of this type also occur; for in-

(71)

stance, the breadth at the foot is half of the maximum breadth, and the breadth at the neck is half that at the lip.

In a like manner, if the distances between the horizontal lines of the characteristic network are comparable in that relationships of equality or of a ratio two to one occur, the fact is appreciated by the eye. For example, in the same illustrative vase the successive horizontal lines are separated by distances in the ratios of $1:1:2:1:1$ as we pass upwards, and these relationships are intuitively enjoyed.

Likewise an interrelationship of equality or of a ratio of two to one between these vertical and horizontal distances may be effective. Thus the part of the vase of Figure 12 below the line of greatest breadth is obviously enclosed in a square.

Moreover, the same vase shows that the perpendicularity and parallelism of characteristic tangents, and also the passing of these tangents through a center of the vase, are felt. By a 'center' is meant a point of the axis of symmetry lying on a horizontal line of the characteristic network of maximum or minimum breadth.

Thus we are led to the following four types of elements of order: relationships between the horizontal distances of the characteristic network, corresponding to the element of order H; relationships between the vertical distances, corresponding to V; interrelations between the two, corresponding to HV; and relations involving the characteristic tangents, corresponding to T.

Aside from these formal elements of order, H, V, HV, and T, I can find none which seem to be generally appreciated and which involve only the form of the vase.

In the empathetic theory of Lipps, the vase is thought of solely as a container, that is as a vessel which is pressed downward by the liquid within it, and which resists this pressure by its elastic strength. This fact seems to be irrelevant as an aesthetic factor, since the empathetic qualities of different shapes are not to be compared aesthetically.

The theory of Hambidge concerns itself mainly with geometric ratios derived from irrational numbers. Such ratios may have been used in actual construction, but in my opinion they cannot be appreciated by the eye.

VASES

6. The Problem of Regularity of Contour

The above considerations justify us in proceeding to the explicit formulation of a definition of aesthetic measure as soon as we have formulated requirements of regularity of contour and of suitability. It will then be possible to limit attention to those vases which meet these requirements, and to develop an appropriate definition of aesthetic measure for them.

In attacking the problem of regularity of contour we shall regard the characteristic points and tangents as assigned. It will be found on experiment that there is little latitude in the choice of a curve of contour, once these characteristics are fixed. Any modification will impair the suavity of the best possible contour, which we shall define presently in precise terms.

When any set of characteristic points and tangents is being considered, it is taken for granted that it is a *possible* form from the geometrical point of view. This is not always necessarily the case. For example, it would not be geometrically possible to have two adjacent characteristic points at which the characteristic tangents are vertical, that is parallel to the axis of the vase. In fact between two characteristic points with vertical tangents there must lie at least one point of inflection or corner.

7. Curvature of Contour

The curvature of any circle varies inversely with the length of its radius, and therefore is regarded as measured by the reciprocal of the radius, taken in given units of length. For example, a circle of radius 2 centimeters has a curvature $\frac{1}{2}$ throughout, if the centimeter is the unit of length.

Now in any curve of variable curvature there is at every point a tangent circle which fits the curve most closely near the point of tangency. It is natural to regard the reciprocal of the radius of this 'osculating circle' as measuring the curvature at the point. The radius of the osculating circle is called the 'radius of curvature.' We shall not stop to formulate a more precise definition of these terms, such as is contained in any elementary book on the differential calculus. In tacitly assuming that the contour lines of a vase possess measurable curvature, we have idealized the actual situation of course.

It is obvious that the problem of regularity is a question concerning the curvature of the contour lines of the vase. In general terms, the contour corresponding to the prescribed characteristic points and tangents should be one of as small curvature as is possible.

8. Requirements for Regularity of Contour

The following further consideration of the requirements for regularity of contour is to be regarded as tentative. It is clear that such requirements can never be given any very satisfactory formulation.

Consider a convex curve made up of arcs of circles of different radii, tangent to one another at their common end points. Evidently the impression produced is not that of a single unified curve, especially if the radii are alternately larger and smaller.

A first obvious requirement, therefore, is that the curvature varies gradually (that is, continuously) along the curve and oscillates as few times as possible in view of the prescribed characteristic points and tangents. In particular the curvature should not oscillate more than once on any arc of the contour not containing a point of inflection. By inspection of various vase forms like those shown later, it is found that this condition is satisfied in practice.

A second like requirement is that the maximum rate of change of curvature be as small as possible along the contour. This condition eliminates both unnecessarily large curvatures along the contour and unnecessarily rapid changes in curvature.

Although the strict application of these two conditions would be very difficult, still they may be regarded as substantially satisfied when it is not feasible to modify the curve so as to diminish either the greatest curvature or the rapidity of change of curvature. The curves of contour actually employed will be found not to permit of such modification.

The eye can follow with ease curves meeting these two requirements, just because of the small curvature and its small rate of change.

9. Conventional and Utilitarian Requirements

There are certain further requirements that are imposed because of the original function of the vase as a useful container which should be of substantial capacity, stable in horizontal position, easy to handle, empty, and move about. Despite the fact that vases are usually employed for decorative purposes, they still suggest their original utilitarian function. In consequence, certain undesirable connotations arise when the vase is felt to violate the requirements which the original function of the vase imposed.

Some of these may be formulated arbitrarily as follows: (1) The breadth at the base should be at least one eighth of the height; (2) The maximum breadth should be attained but once, between the ends of the vase. This breadth should be at least one quarter of the height; (3) The minimum breadth should be attained in the upper half of the vase and should be at least one eighth the height; (4) The height of the vase should be at least as great as its breadth; (5) Rectilinear parts of the contour are only permitted at the ends, of lengths not more than one fourth of the breadth there, and directed so that the breadth is not decreasing toward the end in question; (6) The characteristic tangents at the foot should be inclined at an angle of at least 45°.

These conditions are satisfied in almost all actual vases, and their genesis is easily conjectured. Evidently (1) ensures some degree of stability. The condition (2) ensures that the vase can be easily grasped where it is broadest, and that it is broad enough to have fair capacity as a container. If (3) is satisfied, the vase will have a neck in its upper half by which it may be lifted while it remains in equilibrium; this requirement shows why a vase form cannot in general be inverted and still remain satisfactory as such. Unless the vase is at least as high as it is broad, as required by (4), it appears more like a pot than a vase, and its contents are poured out with difficulty; furthermore, in a squat vase the curve of the contour is relatively inconspicuous. Protective rectilinear parts to the extent (5) are felt to be allowable because the lip and foot are especially liable to injury. If (6) is not satisfied, the vase bulges out so rapidly near the exposed foot that it is not sufficiently protected from injury there.

There are other practical requirements, dependent on the material employed. For example, corners in the contour are more practicable in a metal or lacquer vase than in one of porcelain.

AESTHETIC MEASURE

We shall suppose henceforth that all the stated requirements of regularity of contour, together with these further conventional and utilitarian requirements, are satisfied.

10. On the Interpretation of Vase Form

There are difficulties in passing from an actual vase form to the corresponding symbol.

A vase can be looked at from various positions, each furnishing a somewhat different symbol. The location of the characteristic points and of the directions of the characteristic tangents cannot be determined with much accuracy, as the photographs of actual vases reveal. Furthermore, there are often slight asymmetries, especially in ancient vases, which render the task even more difficult.

But, besides these practical difficulties, there is one of another type in connection with the interpretation of a vase form. A vase is usually slightly rounded at the lip; this conventional type of termination of the contour avoids the suggestion of sharpness. How is this rounding to be interpreted in determining the symbol of the vase? In general we shall either ignore it, or simplify it as far as possible.

11. The Complexity C of Vase Form

With these preliminary definitions and requirements in mind, it is possible to formulate a tentative theory of the aesthetic measure M of vases. We have already noted that it is only the characteristic points of the contour of the vase upon which the eye can rest as it follows the contour lines. It seems natural to consider the number of these characteristic points to be a suitable measure of the complexity C of a vase form. With this definition of C, the complexity of ordinary vases ranges between 6 and 20.

12. The Order O of Vase Form

In section 5 the appreciable elements of order in vase form were analyzed as of types H, V, HV, and T. In order to define O it only remains to assign indices to the types of elements of order there specified. The precise definitions are given below.

VASES

H

The eye appreciates the ratios 1 : 1 and 2 : 1 between the horizontal distances of the characteristic network. The element of order H will be defined as the number of *independent* relations of this kind.

Suppose, for instance, that $a = b$ and $b = c$ are such relations. Then the relation $a = c$ would not be counted since it is not independent of the other two relations. This limitation is made in accordance with the general principle that only independent elements of order are to be counted (Chapter I, section 18).

Moreover we agree never to count H as more than 4, since this element of order has a limited value no matter how often repeated.

V

Similarly the eye can appreciate the ratios 1 : 1 and 2 : 1 in the vertical distances between the horizontal lines of the characteristic network. The element V is the number of independent relations of this kind. The element V will never be counted as more than 4.

HV

There may be ratios 1 : 1 or 2 : 1 between these horizontal and vertical distances of the network. These too are appreciated by the eye. The element HV is the number of such relations, mathematically independent of each other and of those already listed in H and V, which are of one of the following restricted types: the horizontal or the vertical distance in question is the breadth or height of the vase; the horizontal and vertical distances involved are adjacent in the network.

The element HV will never be counted as more than 2, since interrelations of this type are much more uncertain in their aesthetic effect, even when thus restricted.

T

The following appreciable relationships are counted in T, in so far as independent of the relations H, V, HV, and of one another, but only up to 4 in any case: (1) perpendicularities of characteristic tangents; (2) parallelisms of characteristic tangents in other than the vertical direction; (3) vertical characteristic tangents at end or corner points, or at points of inflection; (4) characteristic tangents or normals which pass through an adjacent center of the vase.

13. The Aesthetic Measure M of Vase Form

The formula proposed is

$$M = \frac{O}{C} = \frac{H+V+HV+T}{C}$$

where the meaning of each term has been defined above, and may be briefly summarized as follows: C is the number of characteristic points in the symbol of the vase; H (not exceeding 4) is the number of independent ratios 1 : 1 and 2 : 1 in the horizontal distances of the characteristic network; V (not exceeding 4) is the number of independent ratios 1 : 1 and 2 : 1 in the vertical distances; HV (not exceeding 2) is the number of ratios 1 : 1 and 2 : 1 between the horizontal and vertical distances of the network, involving the total breadth or height or adjacent horizontal and vertical distances, which are independent of one another and of the relationships H and V; T (not exceeding 4) is the number of perpendicularities and parallelisms of characteristic tangents, together with the number of vertical tangents at the end or corner points and at points of inflection, and the number of characteristic tangents and normals through an adjacent center, in so far as all these are independent of the relations H, V, HV, and of one another.

14. Application to Some Chinese Vases

The excellence of Chinese vases and other Chinese pottery is well known. Says Hobson * "But nowhere, perhaps, is the supremacy of the Chinese so marked as in the ceramic art. The satisfying shapes of the T'ang pottery, the subtly refined monochrome porcelain of the Sung, and the gorgeous Ming three-color wares, are things unrivalled." Furthermore the Chinese have used the purely symmetric form of the vase in a large variety of interesting shapes. By contrast, Greek vases are generally asymmetric in that they possess handles and other elaborations of form, and are decorated pictorially. Thus, despite their great beauty, they present an extremely complex aesthetic problem, to which our treatment of symmetric vase form *per se* is scarcely applicable.

* *Chinese Art*, New York (1927), p. 14.

VASES

In order to test out the theory formulated above, which was derived mainly by the process of introspective analysis, I selected for examination the photographs of the first eight symmetric vases which appear in a recent work of Hobson (*loc. cit.*). These are numbered VII, VIII, XIII, XV, XXI, XXII, XXIII, and XXV. Among them I have included one 'jar' and one 'bottle' (Nos. XV and XXI) which are of essentially vase form.

The outlines of these vases are given in Plates XVI and XVII, opposite the next page. The symbols are arranged in order of decreasing aesthetic measure, with the number of each vase and the corresponding measure M underneath. The reader who wishes to verify the facts in the case for himself is advised to consult the actual photographs and measure them.

It should be stated that only those elements of order were considered to occur for which ordinary measurement by means of a ruler indicated no appreciable deviation.

In my opinion the large number of these elements of order H, V, HV, and T which are present cannot be explained as accidental, because vases taken at random contain very few such relationships. It seems to me probable that, more or less intuitively, forms of vases were gradually adopted which involved the types of relation recognized by our theory.

15. A Detailed Analysis

We proceed next to the detailed analysis of these eight vases. This will help the reader to decide for himself whether or not the individual elements of order, to which aesthetic importance is attributed by the theory, are actually effective for him.

The following notations will be used in the specification of horizontal distances of the vases:

f = breadth at *foot* of base.

bf, tf = breadths at *b*ottom and *t*op of *f*oot respectively if there is a rectilinear foot.

m = *m*aximum breadth.

n = breadth of *n*eck.

l = breadth at *l*ip.

bl, tl = breadths at *b*ottom and *t*op of *l*ip respectively.

c = breadth at *c*orner.

i = breadth at *i*nflection points.

bi, ti = breadths at *b*ottom and *t*op *i*nflection points respectively.

In the specification of vertical distances, a corresponding double notation is employed, as for example f–l to designate the distance from the foot (f) to the lip (l).

XV

This beautiful Ming jar with $C = 10$ possesses elements of order as follows:

$H = 3 : f = \frac{1}{2}m = i;\ bl = tl$

The breadths at the foot and at the points of inflection are both equal to half the maximum breadth. Furthermore the breadth is constant along the rectilinear lip at the top.

$V = 2 : \frac{1}{2}(f-m) = m-bl;\ i-bl = bl-tl$

The height from the line of maximum breadth to the bottom of the lip is half the height from the foot to this line.

The eye can interpret the vertical distance between the points of inflection and the lower edge of the lip as equal to the height of the lip, and this is done in the interests of economy of effort although the visual measurement of the short distances involved is necessarily inexact.

$HV = 1 : m = f-m$

The breadth of the vase equals the height from the foot to the line of maximum breadth.

$T = 2$. The tangent at the foot is perpendicular to the tangent at the point of inflection. Moreover the eye can interpret the direction in which the curvilinear part joins the vertical lip as being vertical. Both of these elements of order count in T.

We conclude then that O is 8 for this vase and hence that its aesthetic measure M is 8/10 or .80. This is the best of the eight vases considered, in the technical sense of aesthetic measure.

The reader is advised to take up these eight elements of order individually and ascertain to what extent he finds them of value in the aesthetic organization of the vase. The colored Plate XVIII* opposite page 82

* Mr. Stuart Bruce has been so kind as to embody the vase forms as they appear in the four colored Plates. The coloring and decoration are those of his choosing.

PLATE XVI

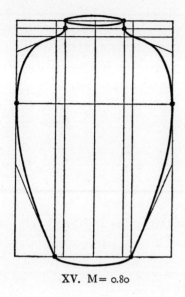

XV. M = 0.80

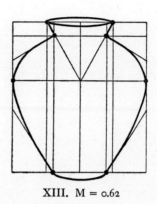

XIII. M = 0.62

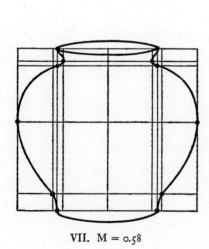

VII. M = 0.58

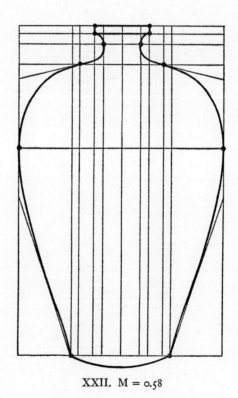

XXII. M = 0.58

AESTHETIC MEASURES OF 8 CHINESE VASES (XV, XIII, VII, XXII)

PLATE XVII

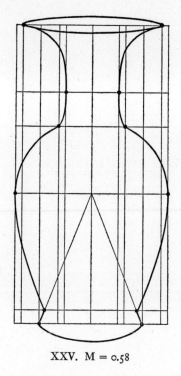

XXV. M = 0.58

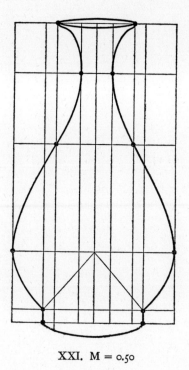

XXI. M = 0.50

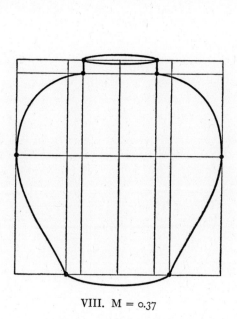

VIII. M = 0.37

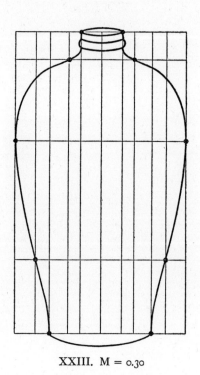

XXIII. M = 0.30

Aesthetic Measures of 8 Chinese Vases (XXV, XXI, VIII, XXIII)

VASES

gives an idealization of this particular vase form which may be of use in this connection. For me, each of these eight elements of order seems to be of positive value.

In this vase and in one or two others of the eight selected vases, there was a small asymmetry in the two contour lines, as far as could be determined from the photograph. Such defects have been eliminated in the contour drawings in what seemed to be the simplest manner.

XIII

In this simple Sung vase, C is only 8. Furthermore we find immediately the following elements of order:

$H = 1 : f = n$
$V = 1 : \frac{1}{2}(f-m) = m-n$
$HV = 1 : m = f-n$

As a matter of fact the height from the foot to the neck is slightly less (about two per cent less) than the breadth. However, this difference is overlooked by the eye, particularly because vertical distances are systematically overestimated in relation to horizontal distances.

$T = 2$. The line of the rectilinear part of the lip passes through the center, and the tangent direction at the neck is perpendicular to the direction of this line.

Thus O is 5 and M is 5/8 or .62 for this vase.

VII

This T'ang vase with $C = 12$ is less typical as a vase form than the 'jar' XV. In our interpretation of this form we regard the neck as flaring out to the top in a definite direction, thus disregarding, as usual, the rounding at the lip. The elements of order are as follows:

$H = 3 : f = ti = l; \frac{1}{2}m = n$
$V = 2 : f-bi = ti-l; bi-m = m-l$
$HV = 1 : \frac{1}{2}m = f-m$

$T = 1$. The tangent at the foot is vertical.

Thus there are 7 elements of order in all, so that O is 7, and hence $M = 7/12$ or .58.

Because of the squatness of this vase, with the height actually less than the breadth, there is a feeling that the form is not appropriate for a vase as such. We see then to what a degree the general preliminary qualification that a form be suitable, influences the aesthetic judgment as to pure form.

According to the theory, the form of this vase would be improved if the narrowest part of the neck (visually) were half way between the upper inflection points and the lip. This improvement could be effected by making the lip somewhat less rounded. With this modification, the aesthetic measure would be increased to $3/4 = .75$. On the other hand the lip would then become distinctly sharp, and an unpleasant connotation would be thereby introduced.

XXII

For this decorated Sung vase, C is 12 and O is 7 as follows:

$H = 3 : f = \frac{1}{2}m; \; i = bl = tl$

$V = 2 : \frac{1}{2}(f-m) = m-n; \; \frac{1}{2}(i-n) = n-bl$

$HV = 1 : m = f-m$

$T = 1$. The characteristic tangents at the inflection points and the base are perpendicular.

The fact that the rectilinear edge of the lip is vertical does not count since this relation is not independent of the element of order $bl = tl$ already enumerated in H.

Hence the aesthetic measure M is $7/12$ or $.58$ as before.

A slight asymmetry has been removed by modification of the right contour near the foot. If the left contour near the foot had been modified instead, the interpretation would have been more complicated.

XXV

This Sung vase is of an unusual but graceful shape with a corner at the bottom of the neck. The complexity C is taken as 12, and the elements of order O are as follows:

$H = 1 : \frac{1}{2}bf = n$

$V = 3 : tf-m = m-n; \; \frac{1}{2}(m-c) = c-n = \frac{1}{2}(n-l)$

PLATE XVIII

ILLUSTRATIVE VASE FORM

VASES

Otherwise stated, aside from the rectilinear part of the foot, the successive horizontal lines through characteristic points divide the height of the vase in the ratios $3/2 : 1 : 1/2 : 1$.

$HV = 2 : bf = \frac{1}{2}(bf-n); \; c = m-c, \; l = \frac{1}{2}(b-l)$

We note that there are three elements of order HV which are only counted as 2 in HV, in accordance with the theory.

$T = 1$. The direction of the rectilinear part of the foot passes through the center of the vase.

Hence O is 7 and M is $7/12$ or .58, as in the two preceding vases.

XXI

The rounding at the lip in this Sung bottle is ignored in our interpretation so that C is 12. The elements of order O are as follows:

$H = 2 : bf = tf; \; i = l$

$V = 2 : bf-m = i-n; \; \frac{1}{2}(m-i) = n-l$

$HV = 1 : m = tf-i$

$T = 1$. The inclined normal at the top of the foot passes through the center of the vase.

Hence O is 6 and M is $6/12$ or .50.

VIII

In this decorated T'ang vase, C is only 8. Furthermore we have the following elements of order O:

$H = 2 : f = \frac{1}{2}m; \; bl = tl$

$V = 0$

$HV = 1 : m = f-tl$

$T = 0$

Hence we find that O is 3 and M is $3/8$ or .37.

It will be observed that the uncertain points of inflection near the base are ignored in this interpretation.

XXIII

This is a Sung vase of porcelain with carved designs.

By a suggested simplification of the neck we may regard C as only 10, as is indicated in the tracing.

AESTHETIC MEASURE

On examination, elements of order O are found only as follows:

$H = 1 : \frac{1}{2}(bi) = ti$

$V = 0$

$HV = 0$

$T = 2$. The characteristic tangents at the foot and lip are vertical.

In consequence the aesthetic measure M is only .30 in this case.

16. Three Experimental Vases

As an interesting test of my theory, I presented a single experimental vase form at Bologna in 1928 (see the Preface). This appears in the colored Plate XIX seen opposite. One finds immediately the following count:

$C = 10$

$H = 4 : f = \frac{1}{2}m = i = 2n = l$

$V = 3 : \frac{1}{2}(f-m) = m-i = 2(i-n) = n-l$

$HV = 1 : m = f-m$

$T = 2$. The characteristic tangents at the top and at the inflection points pass through a center of the vase.

Hence the aesthetic measure M is 10/10 or 1, which is higher than that of any of the eight Chinese vases analyzed above.

A second experimental form is provided by the obvious idealization of the well known Chinese vase with cover or tea jar, in the direction indicated by our theory. The vase form thus obtained (colored Plate XX) is less rotund and capacious as a container than the usual vase of this type, and, to the extent that this is the case, is felt to be less satisfactory even if more elegant. Evidently connotations of utilitarian origin enter into this judgment, such as lie outside the purely formal field to which we restrict attention. However, it will be found, I think, that the deviation of the usual form from this ideal form is so slight that the usual form seems to possess the same elements of order.

The analysis of this form yields the following count:

$C = 7$

$H = 2 : f = \frac{1}{2}m = n$

$V = 1 : \frac{1}{2}(f-m) = m-n$

$HV = 2 : m = b-n; \frac{1}{2}n = n-l$

PLATE XIX

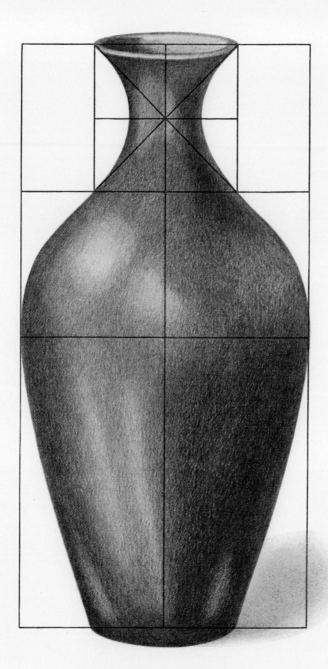

FIRST EXPERIMENTAL VASE FORM

PLATE XX

Second Experimental Vase Form

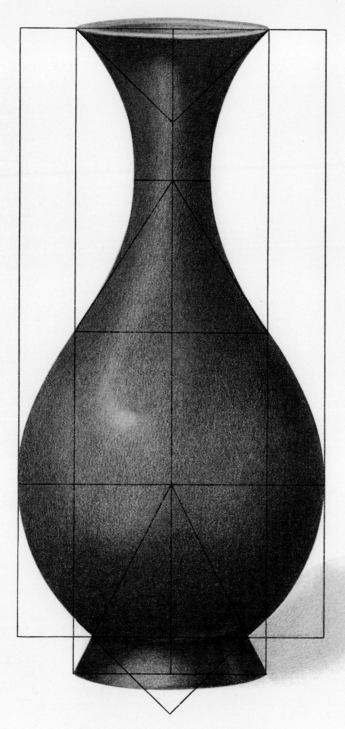

PLATE XXI

Third Experimental Vase Form

VASES

$T = 2$. The characteristic tangents at the bottom of the cover are vertical and horizontal.

Hence the aesthetic measure is 7/7 or 1 in this case also.

Finally the vase in the colored Plate XXI is analyzed as follows:

$C = 12$
$H = 4 : bf = i = l; tf = \frac{1}{2}m = 2n$
$V = 3 : tf-m = m-i = i-n = n-l$
$HV = 2 : 2n = n-l; f = f-m$

$T = 4$. The characteristic tangent at the inflection points and the line of the rectilinear foot pass through an adjacent center. The characteristic tangents at the lip and those above the foot are perpendicular.

This vase has accordingly an even greater aesthetic measure of 13/12 or 1.08.

It may also be mentioned that the vase whose visual contour appears in Figure 12 provides another experimental illustration with $C = 12$, $H = 4$, $V = 4$, $HV = 1$, and $T = 4$, so that here also $M = 13/12$ or 1.08.

17. General Conclusions

My estimate of the actual significance of the above theory of vase form is that it is far more uncertain than any of the other similar theories presented herein, but that it does possess a limited validity as follows:

The aesthetic factors in vase form appear to divide themselves conveniently into three categories: (1) regularity of outline; (2) utilitarian and conventional requirements; (3) definite geometric relationships between the dimensions involved. The first type of factor is purely formal but difficult of precise formulation (see section 8). The second type is evidently connotative in nature and is beyond precise analysis. The third type is formal and is the only one of which our theory takes explicit account, despite the fact that the other two types are of co-ordinate importance.

Now among all the infinitely many conceivable elements of the third type, we have arbitrarily selected only those which involve simple relations between certain prominent horizontal and vertical distances, and between certain tangential directions. It is plausible that these elements of order should be felt, inasmuch as the eye constantly and intuitively compares such distances and directions in everyday experience, and these elements fix the otherwise indeterminate contour curves of vase form. Nevertheless it is entirely possible that other relations are equally felt.

It is our maximum claim, then, not that aesthetic measure as here defined yields a satisfactory aesthetic measure of vase form, but rather that when all requirements of regularity of outline as well as utilitarian and conventional requirements are satisfied, vase forms are in general good in proportion as relations H, V, HV, and T are present.

It is an interesting question as to how the explicit knowledge that these simple relations exist in a given case affects the aesthetic judgment. In my opinion the effect is slightly adverse, for, in all fields of art, it is the intuitively felt relationships which are the most enjoyed.

CHAPTER V

THE DIATONIC CHORDS

1. The Problem of Musical Form

MODERN Western music must be regarded as an unparalleled artistic achievement. In every age and civilization music in simple forms has played an important part. But only in Europe since the Renaissance has it broken the bounds of homophonic song and thereby acquired an almost transcendental expressive power.

From the formal point of view, Western music stands pre-eminent by virtue of its purity and its extraordinary degree of development. In poetry there are formal elements which can be isolated and analyzed. But in poetry the meaning is of such dominant importance and so completely eludes formal analysis, that the field of poetry is not pure in the same sense. Similarly it is obvious that other aesthetic fields are inferior to music, either in purity, or else in degree of development as in the case of polygonal form. For, in the case of music, we have a succession of musical sounds, characterized by pitch and time, replete with relationships and devoid of obvious connotations. Furthermore this music has a deep and almost universal appeal.

All of these considerations indicate that Western music should afford an ideal and perhaps crucial field for the application of our general theory. In fact the complexity, C, of a given musical structure should be readily measurable; the elements of order, O, which are appreciated by the ear, should be determinable; and then we should expect to find that, with suitable assignment of indices to these elements, the aesthetic measure, M, would be given as the density of the elements of order in the musical structure, that is as O/C.

Music owes much of its aesthetic importance to its peculiar emotional effect. This attribute is readily understood. The human voice is a primary means of human expression, and at the same time is a musical instrument. All of us become accustomed to its musical tones, and simultaneously learn to appreciate that which the voice expresses. Thus musical tones have become intimately associated with emotional feeling.

In this attribute lies not only the most subtle source of musical appeal, but also the final limitation of any purely formal theory. For, according to our general theory, music will be most effective when it unites surpassing beauty of form with effective suggestion of emotional utterance.

This elusive power of suggestion plays a relatively unimportant part in the simpler forms of pure music such as we shall consider. At the same time it should not be forgotten that, in the last analysis, one musical work is likely to appeal to us more than another, not because of any superiority of abstract form, but rather because of some connotative suggestion of this kind.

The present and following two chapters present an attempt to analyze explicitly some of the simpler formal elements of order in music, and thus to obtain a suitable definition of aesthetic measure. As far as I am aware, no systematic attempt at a rationalized explanation of musical effects, and in particular of melody, has been made previously. In fact the usual treatments of musical form either state empirical rules of known value or fall into the opposite extreme of literary discourse.

At every stage of musical development certain specific limitations in the use of available musical forms have been observed. These forms have, however, undergone a process of gradual elaboration, and new forms have been established. The best composers of each period have been amazingly successful in discovering the latent possibilities nearest at hand and exploiting them thoroughly. In this connection it is significant that composers seldom achieve much success exclusively along lines employed by their predecessors.

Since the appreciation of relationships of order among musical notes is thus continually expanding and developing, it would be highly absurd to try to formulate a definitive theory of aesthetic measure, valid for the music of the future as well as of the past.

Consequently the problem of musical form, in order to be precise, must involve a definite allowance of musical means. A satisfactory theoretic solution of this problem should account for the large body of simple classical music in which many limitations are observed. If such a solution

THE DIATONIC CHORDS

were successful, it would then be anticipated that, by suitable extension of the underlying principles, the same kind of explanation could be applied to more elaborate musical forms.

Thus the main problem of musical form may be stated as follows:

With a given allowance of musical means, to determine the extent to which the relations of order among the notes of a musical composition constitute the effective basis of musical enjoyment.

The principal claim of the theory of aesthetic measure, by the aid of which we attack this very difficult problem, is that the aesthetic effect is essentially a summational one, due to the presence of various elements of order. The elements which we shall take account of are simple and in large part obvious.

2. Harmony as an Aesthetic Factor

It has long been customary to reduce music to harmony, melody, and rhythm. For the present we shall accept this conventional division of the subject.

Of the three factors harmony is the most elementary, for it is concerned with the aesthetic effect of a complex musical sound (a chord), and of two or even several such sounds heard in succession.

Despite the fact that harmonic intervals such as octaves and perfect fifths between musical notes, are basic in the older homophonic music, it can scarcely be claimed that harmony plays a rôle in it, although melody and rhythm obviously do. However, with the advent of polyphonic music, harmony became important; and the modern ear has grown so accustomed to it, that a simple mental harmonization is automatically effected when any tune is heard.

In order to formulate precisely the problem of harmony, it is desirable that certain facts of fundamental importance be recalled.

3. Consonance and Dissonance

Musical sounds are differentiated from all others, and in particular from noises, by the fact that the vibrations of the air, which impinge upon the ear and produce the sensation of sound, are periodic. The number of vibrations per second determines the pitch of the musical sound; thus the note which we shall designate as middle C has a pitch with vibration frequency of about 258 per second.

The fact that pitch is numerically measurable was known to the early Greek philosopher Pythagoras who observed that if the length of a musical string be divided in the ratio of 1 to 2, then the note of the shorter string is an octave higher. Thus the unmistakable relationship of a note and the note an octave higher, which makes the second note seem to the ear almost identical with the first, is associated with the simple ratio of 2 to 1 between the lengths of the strings which produce the two notes. Furthermore, this relationship is agreeable to the ear, and is properly called consonance. From the modern point of view the fact involved is more conveniently expressed in the form that the pitch numbers or frequencies of a note and of its octave are in the ratio of 1 to 2; for, when the length of a musical string (under a specified tension) is halved, its frequency of vibration is doubled, as is that of the attendant vibration of the air.

Similarly it appeared that if the length of a musical string be divided in the ratio of 2 to 3, the higher note is agreeably related to the lower, being a perfect fifth higher, in the usual terminology. For example, the perfect fifth above C on the pianoforte keyboard is G, being fifth from C if we start to count the white keys from C as first. Likewise if the lengths are in the ratio 3 to 4, the higher note is a perfect fourth above the lower, and is agreeably related to it. The perfect fourth above C is F on the pianoforte keyboard. The frequency ratios of the perfect fifth and perfect fourth to the lower note are of course 3/2 and 4/3 respectively.

Thus Pythagoras discovered that consonant musical notes are in general produced when the lengths of the corresponding musical strings are in the ratio of small integers; this fact signifies to us that the frequencies are in the reciprocal ratios. His observation was essentially that a certain element of order (a simple vibration ratio) is correlated with a certain aesthetic value (the consonance of two notes), and so may be regarded as a primitive illustration of our general thesis.

Two notes which are consonant may be sounded either simultaneously or in succession with pleasurable effect.

On the other hand, if two musical sounds are heard whose frequencies are not thus related, the effect is that of unpleasant dissonance. If, for

THE DIATONIC CHORDS

instance, we strike the notes C and D on the keyboard, with frequency ratio of 8 to 9 (nearly), the effect is dissonant; it is even more dissonant when we strike B and C for which the frequency ratio is 15 to 16 (nearly).

4. Musical Notes, Intervals, Triads, and Chords

A tuning fork or other resonator produces a *pure* musical note in the sense that the attendant vibratory motion of the air is not only periodic but is 'sinusoidal.' On the other hand, a musical instrument such as the violin or human voice produces a musical note for which the vibration is periodic but not necessarily sinusoidal. Nevertheless it may be experimentally demonstrated that this note can be reproduced, both in pitch and timbre, by exciting simultaneously a number of resonators with the frequency of the given note and exact multiples thereof. Care must be taken to excite each resonator to the right intensity.

These facts show that any musical note is to be regarded as made up of a certain pure fundamental note of the same frequency, and of the first, second, etc. overtones of double, triple, etc. the frequency of the fundamental note. In fact the trained ear can distinguish the various components of a musical note (fundamental note and overtones). For such musical notes, the fundamental note strongly dominates, and the successive overtones are usually of diminishing intensities. The timbre of the notes produced by a musical instrument depends only on the relative intensities of the dominating fundamental note and of the various overtones; these are usually approximately the same throughout the entire range of a well constructed instrument.

Suppose now that two pure musical notes are sounded together, the frequency of neither being an exact multiple of that of the other, although the ratio of their frequencies is expressible in whole numbers. For the sake of definiteness suppose that their frequencies are represented by 2 and 3. If then we consider a pure note of frequency expressed by 1, the first of the two notes is the first overtone of this note, and the second note is its second overtone. Therefore the two notes sounded together can be regarded theoretically as the combination of a fictitious fundamental note and its first two overtones, in which, however, only the overtones actually appear. This will obviously produce a musical sound of entirely different character from a natural musical note in which the fundamental note predominates.

In harmony we deal with musical sounds made up of such combinations of natural musical notes whose vibration frequencies are in the ratios of whole numbers. These are called 'intervals' if two notes enter, and 'triads' if three notes enter. Whenever two or more notes enter, the sound will be called a 'chord.'

The consonance or dissonance of such combinations of musical notes can be readily explained in two ways. We may either agree with Helmholtz that, when the frequencies are in the ratios of small integers, the resulting musical sound is pleasantly consonant because the excitation of the ear is regular in character; in the case of large whole numbers the unpleasant dissonance would then be attributed to the irregularity of the excitation. Or, we may bear in mind that any note and its overtones are co-present in all musical sounds, inclusive of those of the voice, and for that reason are associated. Hence we may expect that any two notes with a common overtone will also be felt to be associated. Consequently we may look upon the feeling of consonance or dissonance as arising according as there is or is not association of this kind.

According to the second type of explanation, when we hear an interval made up of C and G for instance, this will appear to be consonant because the second overtone of C, with frequency expressed by 3 if that of C is 1, is the same as the first overtone or octave of G with frequency expressed by 3/2. On the other hand, if we hear C and D together, the eighth overtone of C is (theoretically) the seventh overtone of D; hence if there is any such association, it is very remote, and this fact would explain the obvious dissonance of the interval of a full tone.

Very probably a combination of the two types of explanation is nearer the truth than either separately.

5. The Natural Diatonic Scale

All musical systems employ some set of simple musical notes of definite pitches. Without such limitation, the ear would be lost in a labyrinth of musical sound. Such a set of notes may be said to form a musical 'scale.' All known scales are much like our own; in particular that of the ancient

THE DIATONIC CHORDS

Greeks was almost the same. The reason for this similarity is not far to seek: any scale may be looked upon as the outcome of an effort to arrive at a set of notes with the closest possible relations of consonance.

On this basis the natural diatonic scale is readily explained, and is highly successful and indeed almost inevitable. All other scales are either contained in it, as the pentatonic scale is, or differ from it slightly, or are further elaborations like the so-called quarter-tone scale.

In fact let us start with an arbitrary note, say C. The notes most closely related to it are its octave c which is the first overtone, and the octave above the perfect fifth G above C, which is the second overtone g. But it is immediately verified that the ear regards any note and its octave as substantially equivalent so far as musical effect is concerned. Thus at the very outset we are led to include in our scale the notes

$$\ldots c_1, \quad g_1, \quad C, \quad G, \quad c, \quad g, \quad c', \quad g', \ldots$$
$$\ldots \tfrac{1}{2}, \quad \tfrac{3}{4}, \quad 1, \quad \tfrac{3}{2}, \quad 2, \quad 3, \quad 4, \quad 6, \ldots$$

where we have written the frequency ratios underneath the corresponding notes. Of course we have already noted that, if the frequency of C is represented by 1, that of G is represented by 3/2. The meaning of the notation used is obvious, e. g. g_1 is an octave below G, g an octave above G, and so on.

Suppose that we inquire next for the note which is related to c as C is to G, that is for the note of which c is the perfect fifth. This will be the perfect fourth F above C, of frequency given by 4/3, since we have the proportion

$$\tfrac{4}{3} : 2 :: 1 : \tfrac{3}{2}.$$

Thus our scale at this stage takes the form

$$C, \quad F, \quad G, \quad c,$$
$$1, \quad \tfrac{4}{3}, \quad \tfrac{3}{2}, \quad 2,$$

where we restrict attention to the compass of a single octave, since the extension to higher and lower octaves is immediate.

The three notes C, G, and F are the so-called 'primary notes' of the scale, and are termed 'tonic,' 'dominant,' and 'subdominant' respectively.

(93)

AESTHETIC MEASURE

The remaining notes are then obtained as follows: The third overtone of C, not yet considered, is two octaves above C with a frequency ratio 4, and so yields nothing new. The fourth overtone e′, with a frequency represented by 5, when reduced two octaves becomes E with frequency represented by 5/4. This note, which is a so-called major third above C, is felt by the ear to be consonant with C. The further notes A and B are related similarly to F and G respectively, i.e. form major thirds with these notes.

Finally D is obtained as the perfect fifth of G reduced one octave. The same process applied to the two other primary notes yields nothing new, since the perfect fifths of C and F are G and c respectively. As thus completed, we have the natural diatonic scale:

$$C, \quad D, \quad E, \quad F, \quad G, \quad A, \quad B, \quad c,$$
$$1, \quad \frac{9}{8}, \quad \frac{5}{4}, \quad \frac{4}{3}, \quad \frac{3}{2}, \quad \frac{5}{3}, \quad \frac{15}{8}, \quad 2.$$

The new notes E, A, B, and D are called the 'mediant,' 'submediant,' 'leading note,' and 'supertonic' respectively.

The harmonic advantage of this scale becomes obvious when it is observed that the scale contains the perfect fifth of any note in it except D and B, the perfect fourth of any note except F and A, the major third of C, F, and G, as well as the octaves of all its notes.

6. THE EQUALLY TEMPERED DIATONIC SCALE

Now the natural diatonic scale as thus constructed has a remarkable property which has led to the equally tempered form of the scale, forming the basis of Western music. The eight notes of the octave in the natural scale are so distributed that, to all practical purposes, the successive notes occur either at the interval of a full tone or of a semitone, with C–D, D–E, F–G, G–A, A–B as full tones and E–F, B–c as semitones. More precisely, five further notes $C^\#, D^\#, F^\#, G^\#, A^\#$ can be intercalated so that all the musical intervals between successive notes

$$C, C^\#, D, D^\#, E, F, F^\#, G, G^\#, A, A^\#, B, c$$

are substantially equal; that is, the frequency ratios of successive notes are all nearly equal. As a matter of experience, it is found that the musical

THE DIATONIC CHORDS

relation of two notes depends only on this frequency ratio; for example, all notes and their octaves, with frequency ratio of 1 to 2, are felt to have the same characteristic relation to one another.

In the equally tempered form of the diatonic scale all of these 12 intervals are made *exactly the same*. When this is done the change from the natural diatonic scale is so slight that under ordinary circumstances the ear does not detect the modification.

As far as the equally tempered diatonic scale is concerned, it presents a disadvantage and an advantage, when no use is made of the intercalated notes. The disadvantage is that the precise harmonic relationships no longer hold except between octaves, although the precise relationships are more satisfying than the approximate ones. The advantage is that all the pairs of notes are felt to be consonant excepting only those which fall at adjacent degrees of the scale (up to octaves) and the pair B–F. In fact these consonant pairs, reduced to closest position within an octave, have essentially the frequency ratios

$$2/1, \quad 3/2, \quad 4/3, \quad 5/4, \quad 6/5,$$

involving the small integral numbers 1 to 6, whereas the others have essentially the frequency ratios 9/8, 16/15, 45/32, involving larger integers, and so are felt to be dissonant.

7. The Chromatic Scale. Tonality. Modulation

The decisive superiority of the equally tempered scale first appears when use is made of all the twelve notes of the octave. This augmented 'chromatic scale' has the extraordinary property that any note of it may serve as generating tonic of an equally tempered diatonic scale. Each such note defines a major key. Thus far we have considered only the key of C major, in which the notes are those of the diatonic scale with C as tonic.

For the comprehension of the structure of any possible key, it is convenient to regard any intercalated note such as $D^{\sharp}$, as derived either by heightening the pitch of D, in which case it is called $D^{\sharp}$ (read D sharp), or by lowering the pitch of E, in which case it is called $E^{\flat}$ (read E flat). The white notes and the black notes on the pianoforte keyboard represent

AESTHETIC MEASURE

respectively the notes of the diatonic scale and the intercalated notes, as indicated in the adjoining series:

$$\begin{array}{cccccc}
C^\sharp & D^\sharp & & F^\sharp & G^\sharp & A^\sharp \\
C \cdot D \cdot & E & F \cdot & G \cdot & A \cdot & B \; c \\
D^\flat & E^\flat & & G^\flat & A^\flat & B^\flat
\end{array}$$

If we start from G of the equally tempered scale as tonic, we obtain the scale of 8 notes,

$$G, A, B, c, d, e, f^\sharp, g,$$

and if we start from F as tonic we obtain the scale of 8 notes

$$F, G, A, B^\flat, c, d, e, f.$$

Thus we have specified the keys of C major, G major, and F major in the chromatic scale, of which C, G, and F respectively form the tonic centers. These illustrations show the convenience of such notations as $F^\sharp$ and $B^\flat$.

The notes of a single key are said to possess a definite 'tonality' since they are derived diatonically from a single tonic.

The requirement of tonality has long been fundamental in Western music, and we shall accept this limitation for the general reasons already stated (section 1). From time to time the key may change in accordance with well defined rules, but this takes place in such a way that the feeling of definite tonality in each changing key is maintained. A change of tonality is called 'modulation.'

We shall restrict attention to music in a single definite tonality in order to avoid undue complication.

8. Major and Minor Modes

A selection of notes within a scale may be termed a 'mode.' Within the chromatic scale we have the 'major mode' defined by any diatonic key. There is also the 'minor mode' in the corresponding key. This is obtained by flatting the third and sixth notes of the diatonic scale. Thus the diatonic scale of C in the minor mode is given by

$$C, D, E^\flat, F, G, A^\flat, B, c.$$

Any composition in the major mode may be transposed to the minor mode in accordance with simple rules. There are also well known variants of the minor mode here described, to which we can only allude in passing.

THE DIATONIC CHORDS

While we shall confine attention exclusively to the major mode, it is necessary to bear in mind that there is a parallel minor mode. The interval C to E is called a 'major third,' and is the fundamental third in the major mode. The corresponding smaller interval in the minor mode is the interval C to E^b, and is called a 'minor third' on that account. Evidently it is equal to the interval from E to G in the ordinary diatonic scale since both are intervals of three steps in the chromatic scale. On the other hand the intervals from the tonic to the fourth, fifth, and octave are the same in either mode. These are the so-called 'perfect' intervals.

9. THE PROBLEM OF HARMONY. THE SINGLE CHORD

We have glanced at some of the well known reasons, psychological and aesthetic, which have led to the adoption in the West of the equally tempered diatonic scale. This scale, with its splendid possibilities of harmonization and modulation, forms the basis on which Western music has been built. With these facts in mind we may proceed to consider further the fundamental problem of harmony in this scale.

The technical problem of harmony is concerned with the determination of the conditions under which the aesthetic effect of a single chord, and of two (or possibly several) sounded in succession, is pleasing. A partial solution is to be found in the usual empirical rules and their exceptions, formulated in treatises on harmony.

Such a purely empirical solution of the problem, if it can be called a solution, fails to satisfy the scientific mind. In fact the ordinary person, by merely hearing without analysis a certain amount of music, learns rapidly to appreciate harmony even in its more complicated forms. Furthermore, this appreciation develops in a characteristic manner, so that if various harmonic passages be heard by different persons, the opinions concerning their comparative merits will be almost coincident. This unanimity concerning harmony, and music generally, must rest on some rational basis. Otherwise we must agree with Gurney* and others who postulate a mystical 'musical faculty,' alone capable of discerning the 'ideal movement' which is music. Such a point of view can scarcely be termed rational.

* *The Power of Sound*, London (1880).

If, however, it can be shown that in harmony, just as in polygonal form for instance, there are simple elements of order, O, having an obvious origin, and that these, on being assigned suitable indices, give rise to a satisfactory aesthetic measure for the chord or succession of chords, the problem of harmony will have been solved in a manner which has some claim to being called scientific. Furthermore, the approach to unanimity in aesthetic judgment will then be explained by the fact that the associations which correspond to these simple elements of order are substantially identical for every one.

The simplest problem of harmony is evidently that of the aesthetic measure of the single diatonic chord, which we deal with in the present chapter.

In a theory of the diatonic chords, the question of the complexity may be laid aside at the outset. This is because a single chord, no matter what its constituents may be, is a unitary fact: the only automatic adjustment involved is an incipient adjustment of the vocal cords to *one* of the constituent notes. Hence the complexity C is to be regarded as invariable, so that the aesthetic measure M is identified with the elements of order O according to the fundamental aesthetic formula: $M = O$. Thus we have only to discover the various elements of order which enter, and their indices, and then to unite them by summation.

10. The Intervals

The intervals of the equally tempered diatonic scale (except for the octave) are not natural intervals, and may be completely characterized by the number of steps of the chromatic scale in passing from the lower note to the higher note. The aesthetic effect of any two intervals involving the same number of steps, as two perfect fifths, is the same, at least provided the intervals are heard in isolation and not against the background of an established tonality.

It is convenient to list at the outset the usual names of these intervals, and all instances of them in the key of C major:

0 steps;	C–C, D–D, etc.	unison (or first)
1 step;	E–F, B–c	semitone (or minor second)
2 steps;	C–D, D–E, F–G, G–A, A–B	full tone (or major second)

3 steps;	D–F, E–G, A–c, B–d	minor third
4 steps;	C–E, F–A, G–B	major third
5 steps;	C–F, D–G, E–A, G–C, A–d, B–e	perfect fourth
6 steps;	F–B	augmented fourth
6 steps;	B–f	diminished fifth
7 steps;	C–G, D–A, E–B, F–c, G–d, A–e	perfect fifth
8 steps;	E–c, A–f, B–g	minor sixth
9 steps;	C–A, D–B, G–e	major sixth
10 steps;	D–c, E–d, G–f, A–g, B–a	minor seventh
11 steps;	C–B, F–e	major seventh
12 steps;	C–c, D–d, etc.	octave

Here we confine attention to intervals formed by pairs of notes within a single octave, but with the understanding that the general effect of an interval is not modified if either note is displaced by an octave. Thus we are led to group together pairs of 'complementary intervals' which together form an octave, as follows: unison and octave; minor second and major seventh; major second and minor seventh; minor third and major sixth; major third and minor sixth; perfect fourth and perfect fifth; augmented fourth and diminished fifth.

Hence there are essentially only seven different types of intervals to be considered, of which the one of 6 steps, F–B, B–f, is to be regarded as peculiar in that it is complementary to itself.

Now, in all probability, a wholly untrained ear would grade these musical intervals according to the degree of consonance of the two constituent notes. In that case the semitone and major seventh would seem to be the most disagreeable intervals because of the harsh dissonance, and the full tone and minor seventh would be felt as somewhat less disagreeable. On the other hand, the octave and the perfect fourth and fifth would appear to be the most agreeable of intervals because of their consonant quality.

But those of us who are accustomed to the diatonic scale are affected somewhat differently. Two cases need to be distinguished: the one in which an interval is heard in isolation, and the other in which it is heard with reference to an established tonality. For the moment we direct attention to the first and simpler case.

Under these circumstances the semitone and major seventh are still the most disagreeable of intervals, and the full tone and minor seventh are somewhat less so. The consonant intervals of the perfect fifth and perfect fourth, and especially of the octave, are adjudged, however, to be somewhat insipid. Perhaps this is the case because so many instances of these intervals occur in the diatonic scale, and the two notes are too closely related.

Thus, among the consonant intervals, it proves to be the major and minor thirds and sixths which are most pleasing to the trained ear, for they are found to possess the requisite degree of consonance, while the two constituent notes do not blend too completely. The thirds are liked best of all. Perhaps this is because the succession of two notes involved is so easily sung.

We have still to consider the interval of 6 steps. This interval of the diminished fifth is definitely dissonant. But it has also the unique property of being complementary to itself; in other words the lower note is related to the upper note just as the upper is to (the octave of) the lower. This property undoubtedly lends a peculiar interest to the interval. We shall assume that its favorable aesthetic effect more than offsets the mild dissonance, and so we shall classify this interval as in the same group with the thirds.

In this way we are led to arrange all possible intervals of the diatonic scale, considered without reference to an established tonality, in the following order of preference: major and minor thirds, diminished fifth or augmented fourth; major and minor sixths; perfect fourth and fifth, unison, octave; major second, minor seventh; minor second, major seventh.

The above partial genetic explanation for the usual order of preference among the intervals is essentially similar to the explanation of all elementary aesthetic phenomena, and does not seem to admit of being carried further to advantage.

It will readily be verified that, when the notes of the diatonic scale are actually held in mind, so that the feeling of tonality is established, this classification of intervals no longer suffices. For instance, in the key of C major, the perfect fifth D–A is definitely less agreeable than the perfect fifth C–G. Under similar circumstances the major second C–D is found to be definitely less agreeable than the major second F–G. In both

THE DIATONIC CHORDS

cases the cause of the difference lies in the fact that the intervals C–G and F–G, whose values are enhanced, are construed by the ear to lie in major chords, namely in the so-called tonic chord C–E–G and dominant 7th chord G–B–d–f respectively.

11. The Major, Minor, and Diminished Triads

Let us continue the classification of chords by proceeding to the triads. Obviously we can take the three notes of such a triad to be in closest position, within the compass of a single octave. Henceforth we shall consider triads and other chords only in relation to an established tonality, say that of C major.

If two of the three notes differ by a semitone or major seventh, there will be marked dissonance, and the triad will be unpleasant.

Even if two of the constituent notes differ by a full tone or minor seventh, the triad will in general be found sufficiently dissonant not to be pleasant. But there are two definite exceptions. Both of the triads D–F–G and F–G–B contain the dissonant interval F–G formed by the subdominant and dominant of the scale. Nevertheless these two dissonant triads are pleasing. However, this is not an intrinsic property of the two triads, for, if we repeat one of them, as D–F–G, in a different position such as E–G–A, the triad is not satisfactory. Evidently the effect observed is due to the fact that these triads are construed in their first position to lie in the dominant 7th chord G–B–d–f mentioned above.

Let us first consider only the types of triads not containing any pair of adjacent dissonant notes. Of course, if we use the octave C–c, then B must be considered as adjacent to C = c in this sense. Thus it is a question of selecting 3 of 7 objects which are arranged in circular order, in such a way that no two of them are adjacent (see the next figure).

Hence the three notes must occur at alternate degrees of the scale, with the consequent possibilities:

(1) C–E–G, F–A–C, G–B–d;
(2) D–F–A, E–G–B, A–c–e;
(3) B–d–f.

Here the notes are written in ascending order.

AESTHETIC MEASURE

These are the so-called 'fundamental' or 'root' positions of the triads. The lowest note will be called the 'root' of the triad, so that we obtain seven such triads, with a root for each degree of the scale. In consequence these triads may be identified by their roots. Thus D–F–A may be called the supertonic triad, etc.

When the root of a triad is not in bass position, the triad is said to be 'inverted.' Thus in a first inversion the lowest note is the first above the root; in a second inversion, the lowest note is the second above the root. For example, G–c–e is a second inversion of the tonic triad.

The tonic, subdominant, and dominant triads in class (1) are the so-

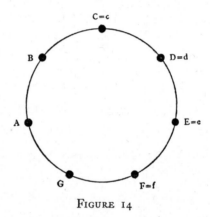

FIGURE 14

called primary or major triads. They are all of identical structure, consisting of a major third followed by a minor third. The intervals of the major third, minor third, and perfect fifth which are involved are all consonant intervals of pleasing type.

Furthermore the root in these cases is properly called the 'generator' of the triad, since the other two notes occur among its overtones of low order. Thus it might be reasonably predicted in advance that these major triads are the best in aesthetic quality, and this fact can be immediately verified. It should be observed that this pleasing quality is independent of any feeling of tonality, and is the same for all three primary triads.

The three triads of class (2), with the supertonic, mediant, and submediant respectively as roots, are the so-called minor triads. They are of identical structure, since each of them consists of a minor third followed by a major third. Thus the same consonant intervals are involved in the

THE DIATONIC CHORDS

minor triads as in the major triads, and this fact might seem to warrant the expectation that the minor triads would be as pleasing as the major triads. They obviously are not, although thoroughly consonant. In fact the third of the chord is not a low overtone of the root, so that the root is not a true generator.

By way of partial compensation, however, there is the familiar parallelism of the major and minor modes, which necessitates that the central tonic triad in the minor mode will be a minor triad. The appropriateness of the terminology of 'major' and 'minor' in this connection is apparent. Moreover, this parallelism renders legitimate the designation of the lowest note as the root of such a minor triad.

Thus we find that the minor triads, when used in the major mode, are not displeasing, but are in general inferior to the major triads.

The diminished triad consists of two superimposed minor thirds of consonant type, and involves the characteristic diminished fifth. Despite its obvious minor quality, the triad produces an interesting and agreeable effect. The root is obviously not in any sense a generator. Because of this fact we shall regard the diminished triad as *without* true 'fundamental position.' In the case of a first inversion when D is in the bass, and the slightly dissonant B, F fall in the upper parts, the chord is scarcely felt as dissonant if these notes are not doubled.

The diminished triad is also used frequently in such a way as to partake of the major character of the dominant 7th chord G–B–d–f already referred to. We shall not consider this to be an intrinsic property of the triad.

All of these triads are consonant, with the exception of the diminished triad in which the root and fifth are dissonant. The remaining triads, which involve at least a full tone or semitone dissonance, are in general not pleasing. They will not be considered further at this point. .

12. The Corresponding Chords

When the constituent notes of any chord whatever are reduced to closest position within the octave, there will always be marked dissonance if two of these notes differ by a semitone or major seventh. Moreover, even if two of these notes differ by a full tone or minor seventh,

AESTHETIC MEASURE

there will be dissonance which is generally unpleasant. Hence, just as in the case of the triads, we are led to consider first those chords in which no two notes fall at adjacent degrees of the scale.

The case in which only one note enters can be passed over without comment, aside from the statement that such a note will be construed by the ear to lie in some definite triad in accordance with definite laws (section 18). If two notes enter, the chord is essentially an interval, and will again be construed as in some definite triad.

When three notes enter, there is one and but one corresponding triad. Consequently every such chord may be assigned to a corresponding triad, whose name it bears. Similarly the chord will be said to be in root or fundamental position in case the bass note is the root of the triad. A like terminology will be applied to the inversions of the chord.

If four or more notes enter into the chord, there will be at least two notes on adjacent degrees of the scale. Hence such chords are dissonant and in general not pleasing.

There are, however, certain exceptional cases. For example, the dominant 7th chord, G–B–d–f, is distinctly agreeable despite the dissonant intervals G–f and B–f. This chord may be regarded as the prototype of the dissonant chords most used in practice. We proceed with the consideration of such chords.

13. THE DOMINANT 7TH AND 9TH CHORDS

The most important of dissonant chords is the 'dominant 7th chord,' formed by a major third, perfect fifth, and minor seventh, placed above the root. In the diatonic scale of C major, this chord appears only in the position G–B–d–f, with the dominant, G, as root. Here there is a single full tone dissonance involved, namely of the root, G, and its seventh, f.

It is easy to explain why the dominant 7th chord is the most agreeable of all possible chords which involve four distinct notes. The second and fourth overtones of the root G are the notes d′ and b′ respectively. This relationship accounts for the pleasing quality of the primary dominant triad G–B–d. Moreover the sixth overtone of G is practically f″, two octaves above the note f. Consequently the dominant 7th chord, thus re-

THE DIATONIC CHORDS

garded, forms a unitary chord of major character of which the root, G, is the true generator. Musical usage has combined with this natural superiority of the dominant 7th chord to establish it firmly.

It will be noted that the dominant 7th chord in its fundamental position is made up of successive thirds. This suggests a further 'dominant 9th chord,' formed by root, its major third, perfect fifth, minor seventh, and major ninth, exemplified in the single position G–B–d–f–a in the diatonic scale of C major. The added ninth almost coincides with the eighth overtone of the root lowered two octaves, so that, like the dominant 7th chord, this chord is a unitary chord of major character.

The dominant 7th and 9th chords retain their characteristic major effect even when inverted, except that in the latter case the dissonant ninth should not fall below or just above the root, or below the third of the chord; the ninth is of course dissonant with both of these notes. If this requirement is not met, the chord is distorted so much as to lose its major character. Such an exceptional dominant 9th chord will be regarded as 'irregular,' in contradistinction to the other regular forms of the chord.

14. The Higher Dominant Chords

By a higher dominant chord we shall mean any dissonant chord containing not only the dominant, G, and certain notes of the dominant 7th chord perhaps, but also at least one of the three remaining mutually consonant notes, A, C, E.

If we proceed from the root G of such a dominant chord upwards by thirds we obtain all of the notes of the octave in the following order:

$$\begin{array}{ccccccc} 1 & 3 & 5 & 7 & 9 & 11 & 13 \\ G, & B, & d, & f, & a, & c', & e'. \end{array}$$

The highest note to appear will be used to characterize the dominant chord. Thus if the seventh or the ninth is the highest note, the chord is called the dominant 7th or 9th chord as the case may be, in agreement with our previous usage; and similarly the chord G–B–f–e′ is a dominant 13th chord.

Such a dominant chord will be said to have the dominant note itself as root, and to be in fundamental position if the root is in the bass. Its successive inversions are calculated on the basis of the complete series displayed above; e.g. F–G–B–e is a third inversion of the dominant 13th chord.

In case the ninth, A, of a dominant 11th or 13th chord is below or just above the root or below the third, or the eleventh, C, is below or just above the third or below the root, or the thirteenth, E, is not at the top of the chord, the characteristic effect of the chord is destroyed. Hence, as in the case of a dominant 9th chord, such a chord will be regarded as regular only if these requirements are met. Otherwise it will be considered to be 'irregular,' and without major character, true root, or corresponding fundamental position.

15. The Derivatives of Dominant Chords

Under certain conditions dissonant chords not containing the dominant note itself are felt to have dominant quality. This happens in virtue of the particular musical context, which is such that either a preceding dominant note is held over in memory, or a subsequent dominant note is present by anticipation. Such a dissonant chord will be termed a 'derivative' of the particular dominant chord suggested.

More precisely, any dissonant chord, not containing the dominant although a note of the dominant 7th chord is dominating (section 21), which becomes a *regular* dominant chord by the addition of the dominant note in some position will be termed a 'derivative' of this dominant chord.

Such a derivative will never be taken to have fundamental position, since the dominant is lacking.

16. Regular and Irregular Chords

All of the fundamental chords and the regular dominant chords have a definite fundamental position. Certain of them, namely the primary chords, the regular dominant chords, and their derivative chords, possess major character.

All other chords of the major mode are said to be 'irregular,' and are never regarded as having fundamental position or major character.

We shall not attempt to consider the chords of the minor mode, despite the fact that the classification proceeds along entirely similar lines.

THE DIATONIC CHORDS

Of particular interest is the 'chord of the diminished seventh,' which in the key of A minor is represented by D–F–G♯–B. The chord of the diminished seventh is only slightly dissonant since it is made up of four equal intervals of a minor third, and involves no dissonances except the diminished fifth. This chord is decidedly agreeable and is very important for harmonization in the minor mode. Because of its structure, the various positions are functionally indistinguishable.

17. Final Classification of Regular Chords

Thus we obtain the following list of regular chords:

    I :   C*–E–G   (tonic)
    ii : D*–F–A   (supertonic)
    iii : E*–G–B   (mediant)
   IV : F*–A–C   (subdominant)
    V :   G*–B–D   (dominant)
   V_7 : G*–B–D–F   (dominant 7th)
   V_9 : G*–B–D–F–A   (dominant 9th)

Irregular if A is below G or B, or one note above G.

   V_{11}: G*–B–D–F–A–C   (dominant 11th)

Irregular if A is below G or B, or one note above G; or if C is below G or B, or one note above B.

   V_{13} : G*–B–D–F–A–C–E

Irregular if A is below G or B, or one note above G; or if C is below G or B, or one note above B; or if E is not at the top of the chord.

    vi : A*–C–E   (submediant)
   vii : B*–D–F   (leading note)

Not dissonant if D is in the bass, and B, F are not doubled.

In each case the Roman numeral refers to the position of the root, which appears first and is starred. The large Roman numerals are used for those regular chords which have major character. The notes which are felt as dissonant in a higher dominant chord are in general those not belonging to the primary, dominant part of it; such a note is dissonant, however, only in case some note with which it is dissonant is present. This is because the higher notes F, A, C, E are felt as dissonant *against* the dominant chord.

All irregular chords may either be thought of as irregular dominant chords, or as formal derivatives of such chords, i.e. as inversions in which the root is absent.

18. Incomplete and Ambiguous Regular Chords

In musical compositions there appear often certain combinations of notes constituted by some but not all of the notes in one of the regular chords. Such an interval or chord will be said to be 'incomplete' in case it contains only two notes, or is part of a higher dominant chord in which the characteristic seventh is absent. Under these circumstances the chord is notably deficient in quality.

Furthermore, in case the notes of an interval are consonant, and can be ascribed to two consonant chords, it will be said to be 'ambiguous.'

Thus the interval C–G in the key of C major is incomplete. It is not ambiguous, however, since it falls only in the tonic chord. On the other hand the interval E–G is not only incomplete but is ambiguous, since it may be ascribed either to the tonic or to the mediant chord.

No other regular chords are to be regarded as ambiguous. For these are assignable to one and only one of the dissonant regular chords.

In the case of an ambiguous consonant chord preceded by a consonant chord, the most important rule of association is readily formulated: Any ambiguous consonant chord is construed to lie in an augmented consonant chord, primary if possible, which contains a maximum number of notes of the preceding chord.

Suppose, for instance, that we have the sequence of two intervals A–c and c–e. If the second interval were heard in isolation it would be construed as in the tonic chord I of course. But, being preceded by A–c, it is clear that the submediant chord vi is the augmented chord determined by the rule. Hence in such a case the interval c–e is construed to lie in the submediant chord. This is in obvious agreement with the facts.

Our reason for specifying the rule of association is to make it clear that ambiguous regular chords are construed in all cases to lie in a definite complete regular chord.

19. The Chord Value C_D

With these preliminaries in hand, we are in a position to define the three elements of order in a chord which we regard as determining its aesthetic measure m.* The chord will be taken in the tonality of a major diatonic key, say C major, and will be regarded as definitely construed in case it is ambiguous (section 18).

The first of these elements will be called chord value and denoted by Cd. It refers to certain obvious attributes of the chord which are not changed when its upper notes are moved up or down by octaves.

We have seen that there is a fundamental division of chords into two types: namely, the regular chords of major character, and all others. Those of the first type have a brightness of tone which makes them generally superior to those of the second. The characteristic differentiation between major and other chords gives rise to the first component in Cd.

Of course this is true only in the tonality of a major key. When a minor mode is used, the superior quality is transferred in part from the major to the minor chords.

Furthermore, other things being equal, the fundamental positions of regular chords are the best. This fact is taken account of by the second component in Cd.

Still another classification separates the possible chords into those which do not involve dissonance, and those which do. A dissonance of a full tone always operates unfavorably, as far as the isolated chord is concerned. If a dissonance of a semitone enters, or of a ninth (sixteenth, etc.), the dissonance is still more unpleasant. The third component of Cd takes account of the effect of dissonance.

Finally, if the chord is incomplete or irregular (section 18), it loses its representative character. The last component of Cd corresponds to this factor.

Hence we are led to define the chord value Cd in the following manner as the sum of four components: a component 1 if the chord is major in character; a component 1 if the chord is in fundamental position; a component -1 if the given chord is dissonant, and -1 in addition if the dissonance of a semitone or ninth is involved; a component -1 if the chord is incomplete or irregular.

It is obvious that certain facts, whose importance has appeared during

* We use the small letter m, instead of M, since the latter is reserved for the aesthetic measure of a sequence of two chords (Chapter VI).

our consideration of chords, are taken account of by this element of order Cd, namely those concerned with major character, fundamental position, dissonance, and incompleteness or irregularity.

20. The Interval Value I

If one examines any melody with a simple harmonization, it will be found that, while other notes of a consonant chord may be lacking, the third above the bass is almost invariably present, and that this is the case to such an extent that the chord seems inferior in case the third above the bass does not appear.

It is interesting to ask for the basis of this feeling. The explanation is perhaps the following: The fundamental triads are the most frequently employed chords, and in particular are best in fundamental position with their roots and generators in the bass. In this event the upper notes are felt to be naturally related to the root. Furthermore such a triad is generally sung in ascending order from the root through the third and fifth. In consequence the third is regarded as the most characteristic note after the root, not only for the major triads but for the minor and diminished triads as well. If the third is actually present, there is a feeling of satisfaction, and, if it is not present, there is dissatisfaction. This feeling is then extended by association to other chords.

The situation is somewhat altered in the case of the dissonant regular chords, as follows. In the case of the higher dominant chords and their derivatives, when the characteristic seventh, F, is in the bass, this note seems to be transferred mentally to the higher parts of the chord; here the note above F plays the rôle of the bass note. If the third above this note occurs in the chord with no dissonant note interposed, the expected third is felt to be present. In the case of irregular chords, if the third above the bass occurs and no dissonant note is interposed, the expected third is again felt to be present.

We shall include all of these possibilities under the heading of the 'expected third.'

Besides the expected third, another interval which is felt pleasantly is that of the diminished fifth. The pleasantness of this dissonant interval has already been alluded to (section 10).

THE DIATONIC CHORDS

With these facts in mind we are led to formulate the following definition:

The interval value I of a chord is the sum of the following two components: 1 if the interval of the expected third appears; 1 if the interval of the diminished fifth appears.

21. THE VALUE D OF THE DOMINATING NOTE

Among the notes of a chord there is a lowest one which is repeated at least as often as any other in the chord. This note will be termed the 'dominating note' for obvious reasons. In the particular case of a chord of four parts, which is of especial interest, the doubled note is the dominating note, unless there are no doubled notes or two pairs of doubled notes; and, then, the dominating note is clearly the bass.

Generally speaking, it is important that the right note of a chord be dominating if the effect is to be as good as possible. It is this last element D of the dominating note that we proceed to consider.

The quality of a chord is enhanced when a consonant primary note is dominating. In fact the brightness of a major chord is thereby increased, while the sombre effect of a minor chord is lessened. If the dominating note is not only such a primary note but is the root of the chord in fundamental position, the best result is obtained of course.

For a diminished triad the only primary note is the fifth, F, which is regarded as dissonant when dominating, even if the third, D, is in the bass; here it is of no advantage to have a dominating primary note.

For the dissonant dominant chords of regular type it has been observed that the primary note G is not felt as a dissonant note of the chord (section 17), and the brightness of these chords is increased if G is dominating.

On the other hand it is very undesirable that a doubled dissonant note — not merely involving the mild dissonance of the diminished fifth in the upper parts — or the leading note, or even the fourth above the bass, be dominating. In fact the leading note strongly suggests the dissonant tonic, and the fourth above the bass is dissonant with the expected third.

All of these facts are taken account of in the following definition of the value D of the dominating note:

AESTHETIC MEASURE

A component 1 is assigned to D if a consonant primary note is dominating, and 1 in addition if it is root and bass of the chord; a component -2 is assigned to D if a dissonant note (not merely a note of the diminished fifth in the upper parts), or the leading note, or the fourth above the bass is dominating and doubled. The value of D is the sum of these components.

22. The Aesthetic Measure m of the Single Chord

The aesthetic measure m of the single chord in a definite tonality will be taken as the sum of the chord value Cd, the interval value I, and the value D of the dominating note:

$$m = Cd + I + D.$$

The definitions of Cd, I, and D may be briefly restated as follows:

Cd is the sum of the following four components: 1 if the chord is major; 1 if it is in fundamental position; -1 if it is dissonant, and -1 in addition if it involves a semitone or ninth dissonance; -1 if it is incomplete or irregular.

I is the sum of the following two components: 1 if the interval of the expected third appears; 1 if the interval of the diminished fifth appears.

D is the sum of the following two components: 1 if a consonant primary note is dominating, and 1 in addition if it is the root and bass of the chord; -2 if a dissonant note (not merely a note of the diminished fifth in the upper parts), or the leading note, or the fourth above the bass is dominating and doubled.

In the application of the formula it is taken for granted that the notes of the chord are not too widely separated. For definiteness, let it be assumed once and for all that the interval between the bass and the first note above it is less than two octaves, and that no other interval between adjacent notes exceeds an octave. If these limits are exceeded, the chord becomes of dubious value, because of its obvious lack of unity.*

It is assumed also that the chord does not consist of only a single note or of the same chord repeated in different octaves.

---

* In general 'close position' is superior to 'extended position,' but we shall not attempt to evaluate this factor.

THE DIATONIC CHORDS

The diagram below not only illustrates the formula, but serves to recall its associative basis according to the general theory.

A cursory inspection of the diagram shows that the measure of no chord can exceed 5, this limit being attained only by the primary and dominant 7th chords in fundamental position, with dominating root.

DIAGRAM OF AESTHETIC MEASURE m OF CHORD

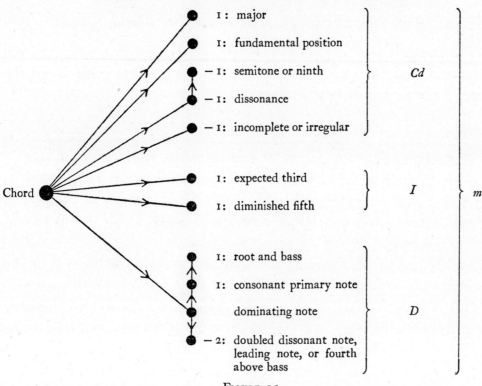

FIGURE 15

The lowest possible measure is -5. We shall find that a chord with measure 4 or 5 may be considered as good, with measure 3 as fair, with measure 2 as passable, and others as in general bad.

Occasionally a chord not even passable must be introduced in a sequence owing to the rigid requirements of the melodic pattern.*

23. THE COMPLETE TRIADS IN ROOT POSITION

It is decidedly interesting to consider the usual rules concerning the single chord in the light of the theory proposed above. Here we shall

* It must be constantly borne in mind that we are referring to music of simple, classical type.

make the natural assumption that, other things being equal, the chords most employed are best in themselves. Frequently used chords will be expected to rate as good or at least fair, with m not less than 3. Likewise, for chords which are used less often, we shall expect to find m equal to 2.

In effecting such a comparison with the usual rules of harmony, we shall limit attention to harmony in four parts, which is the case usually treated, and shall consider these rules as they are formulated by Prout (referred to as P.) in his *Harmony; its Theory and Practice* (thirty-first edition).*

The first general statement is to the effect that "the relative positions of the upper notes of a chord make no difference to its nature, *provided the same note of the chord is in the bass*" (italics as in P., p. 36).

This conclusion is obviously in complete agreement with our formula, in the sense that the values of Cd, I, D, and of their component parts, are not altered by such rearrangement of the upper notes of any regular chord.

Next (P., p. 37) we may quote the statements: "with one exception . . . it is possible to double any of the roots of a triad; but they are by no means all equally good to double. . . . *Double a primary rather than a secondary note.*" This statement is of course directed towards the element D in our formula.

Inasmuch as only the primary triads in fundamental position are being considered by Prout at this stage, we have to inspect the following tabulation of values of m:

$$
\begin{array}{ccc}
\text{C–E–G} & \text{F–A–C} & \text{G–}\overset{*}{\text{B}}\text{–D} \\
5 \ 3 \ 4 & 5 \ 3 \ 4 & 5 \ 1 \ 3
\end{array}
$$

Here the corresponding value of m is listed underneath each possible dominating note in these triads. Furthermore the asterisk above is used to indicate that the corresponding chord is not even 'passable.' It appears then that all of these chords are rated as good when a primary note is dominating, and as fair otherwise, with one exception. This exception is the one referred to, namely, the dominant chord with doubled leading

* We have selected this practical text because it is thoroughly conventional in its treatment, and representative of a crystallized classical point of view.

THE DIATONIC CHORDS

note, which is never allowed "excepting in the repetition of a *sequence*" (P., p. 38).

The second rule concerning doubling is formulated as follows: "*In the root position of a chord, it is seldom good to double the fifth*" (P., p. 37). Evidently this conclusion is justified by reference to the above table in which m is greater by 1 or 2 if the root is doubled, than if the fifth is doubled.

Although these two rules have reference in the first place only to the primary triads in fundamental position, they are represented as equally applicable to the minor triads in fundamental position, while explicit exception is made of the diminished triad vii. But for these minor triads the corresponding tabulation is

$$\begin{array}{ccc} \text{D–F–A} & \text{E–G–}\overset{*}{\text{B}} & \text{A–C–E} \\ 2\ \ 3\ \ 2 & 2\ \ 3\ \ 0 & 2\ \ 3\ \ 2 \end{array}$$

From these results it appears that this extension of the rules is also in accordance with the formula.

Concerning the remaining diminished triad vii, it is stated (P., pp. 37, 38): "a diminished triad is very seldom found in root position, except in a sequence;" "in the chord vii the fifth is the only primary note. Here, however, it cannot be doubled in root position because the fifth of this chord is not a perfect, but a diminished fifth and we shall learn later that it is not generally good to double a dissonant note;" "excepting in the repetition of a *sequence . . . the leading note must never be doubled.*"

The tabulation of this chord, as a derivative of V7, gives

$$\overset{*}{\text{B}}\text{–D–}\overset{*}{\text{F}}$$
$$0\ \ 2\ \ 0$$

These results show that the root and fifth are not to be doubled in any case, and that the chord with doubled third, D, is only rated as passable. This agrees with Prout's statements quoted above.

The fact that the primary or major triads are in general of higher rating than the minor and diminished triads agrees of course with the characterization of the "primary chords, as the most important in the key" (P., p. 49).

(115)

AESTHETIC MEASURE

Thus, so far as the complete fundamental triads in root position are concerned, the ratings given by the formula are in complete agreement with the usual rules.

24. THE INCOMPLETE TRIADS IN ROOT POSITION

In the case of incomplete triads, the ratings of the root positions are as follows:

$$\begin{array}{cccccc}
\text{C–E} & \text{C–G} & \text{F–A} & \text{F–C} & \overset{*}{\text{G–B}} & \overset{*}{\text{G–D}} \\
4\ \ 2 & 3\ \ 2 & 4\ \ 2 & 3\ \ 2 & 4\ \ 0 & 3\ \ 1 \\
\overset{*}{\text{D–F}} & \overset{*\ *}{\text{D–A}} & \overset{*}{\text{E–G}} & \overset{*\ *}{\text{E–B}} & \overset{*}{\text{A–C}} & \overset{*\ *}{\text{A–E}} \\
1\ \ 2 & 0\ \ 0 & 1\ \ 2 & 0\ -2 & 1\ \ 2 & 0\ \ 0
\end{array}$$

Here we have put the major triads in the first line and the minor triads in the second line. The incomplete diminished triad is omitted since it is not passable when it is incomplete. These ratings indicate that the best note to omit is the fifth, in conformity with the rule (P., p. 39): "One note of a triad is sometimes omitted. This is mostly the fifth of the chord — very rarely the third. . . ." It will be noted that all of the triads have passable forms if the fifth is omitted. This latter fact justifies the use of such chords as stated (P., p. 39): "But it not infrequently becomes necessary to omit the fifth, in order to secure a correct progression of the parts."

25. THE FIRST INVERSIONS OF THE TRIADS

Let us turn next to examine the first inversions of the fundamental triads. The tabulation with all three notes present is as follows:

$$\begin{array}{ccc}
\text{E–G–C} & \text{A–C–F} & \overset{*}{\text{B–D–G}} \\
2\ \ 3\ \ 3 & 2\ \ 3\ \ 3 & 0\ \ 2\ \ 3 \\
\overset{*\ *}{\text{F–A–D}} & \overset{*\ *}{\text{G–B–E}} & \overset{*\ *}{\text{C–E–A}} \\
2\ \ 1\ \ 1 & 2\ -1\ 1 & 2\ \ 1\ \ 1 \\
\text{D–F–}\overset{*}{\text{B}} & & \\
3\ \ 2\ \ 1 & &
\end{array}$$

Here we have put the major triads in the first line, the minor triads in the second line, and the diminished triad, taken as a derivative of V_7, in the third line.

(116)

THE DIATONIC CHORDS

Evidently we are led to expect from this tabulation that "all the triads ... can be used in their first inversion" (P., p. 63). Furthermore it clearly remains advisable to double a primary note, excepting in the case of the diminished triad, in accordance with the statement (P., p. 63): "The rule ... to double a primary, rather than a secondary note, applies to first inversions as well as to root positions."

The exceptional case of the diminished triad is treated separately (P., p. 63): "The root, being the leading note, must on no account be doubled except in the repetition of a sequence. As a general rule, *in the first inversion of a diminished triad, the best note to double is the bass note*. But it is not forbidden to double the fifth (the primary note of the chord), if a better progression is obtained thereby." These statements also agree with the tabulated results.

It will be noted, moreover, that for the major triads there is no longer any especial advantage in doubling the primary root instead of the primary fifth, as is indicated in the following statement (P., p. 63): "But the objection to doubling the fifth of the chord no longer holds good."

If we tabulate the incomplete forms of the major and minor triads in their first inversions, we obtain the following:

```
  *      * *      *      * *      * *     * *
 E–G    E–C     A–C    A–F     B–D    B–G
 1 2    0 1     1 2    0 1     –1 1   –2 1

 * *    * *     * *    * *     * *    * *
 F–A    F–D     G–B    G–E     C–E    C–A
 1 0    0 –1    1 –2   0 –1    1 0    0 –1
```

The first inversion of the diminished triad in the incomplete forms D–F and D–B need not be considered as such, since they are construed as in ii and V respectively, according to the rule of section 18.

From this table it appears that the fifth of the chord cannot be omitted in a first inversion except possibly in sequence, although in certain cases the root may be; the third is of course the bass in a first inversion. These conclusions are to be compared with the statement (P., p. 63): "while the fifth is often omitted in a root position, it is hardly ever good to omit it in a first inversion."

In concluding our remarks about the first inversions of the fundamental triads, it may be observed that our formula contains a positive compo-

AESTHETIC MEASURE

nent corresponding to the fundamental position of a regular chord, and so recognizes a certain natural superiority of the fundamental positions as compared with first inversions.

26. The Second Inversions of the Triads

The advantage of securing a naturally moving bass note in a melody outweighs that of using the chords in their fundamental positions, so that first and higher inversions are often employed.

Concerning the second inversion of a triad it is stated that "its effect is much more unsatisfying than that of either the root position or the first inversion" (P., p. 65). From our point of view, the inferior effect is due to the lack of an expected third above the bass, although the fourth, dissonant with the third, is present. This is the gist of the italicized statement (P., p. 65): "*a fourth with the bass produces the effect of a dissonance.*"

Let us tabulate the second inversions of the triads in complete form:

$$
\begin{array}{ccc}
\overset{*\ \ *}{\text{G–C–E}} & \overset{*\ \ *}{\text{C–F–A}} & \overset{*\ *\ *}{\text{D–G–B}} \\
2\ \ 0\ \ 1 & 2\ \ 0\ \ 1 & 1\ \ 0\ -1 \\
\overset{*\ *\ *}{\text{A–D–F}} & \overset{*\ *\ *}{\text{B–E–G}} & \overset{*\ *\ *}{\text{E–A–C}} \\
0\ -2\ \ 1 & -2\ -2\ \ 1 & 0\ -2\ \ 1 \\
\overset{*\ *\ *}{\text{F–B–D}} & & \\
-1\ -1\ \ 1 & &
\end{array}
$$

Here the diminished triad is construed as a derivative of V_7. It appears, therefore, that only the second inversions of the primary triads in complete form are passable, even if the best note is doubled. This is in agreement with the observation (P., p. 66): "Though it is possible to take any triad in its second inversion, the employment of any but primary triads in this position is extremely rare."

This same tabulation is in entire agreement with the following rule (P., p. 71): "The best note to double in a second inversion is not the root of the chord, even when this is a primary note. . . . The bass note itself is almost always the best note to double; but it is possible, and occasionally even advisable, to double either the third or the root of the chord. In the extremely rare iiic [the second inversion of iii], the bass

THE DIATONIC CHORDS

note, being the leading note, must of course not be doubled; here the third of the chord (the primary note) is, as in other positions of the same chord, the best to double."

27. THE DOMINANT 7TH CHORD

Among the regular chords there remain for consideration all of the dissonant chords, aside from the diminished triad. We consider first the dominant 7th chord. The tabulation for the chord in four parts, complete or incomplete, is as follows:

G–B–D–F	* * G–B–F	* * G–D–F	* G–F
5	5 1 1	3 1 –1	2 –2
B–D–F–G	* * B–F–G		
2	–1 –1 2		
D–F–G–B	* * D–F–G		
2	1 –1 2		
* * F–G–B–D	* * F–G–B	* * * F–G–D	* * F G
2	0 3 0	–2 1 0	–3 0

Here the first line contains the fundamental position, and the three following lines contain the three successive inversions respectively.

The first statement to be considered is the following (P., p. 96): "The chord of the dominant seventh is very frequently found with all its four notes present: but the fifth is sometimes omitted, and in this case the root is doubled. Neither the third nor the seventh of the chord can be doubled. . . ."

This statement is in entire accord with the results derived from our formula and tabulated above. It will be observed that the fundamental position is rated as fair ($m = 3$) when the third is omitted provided the root is doubled, and as passable even if the fifth is also omitted provided the root is tripled. However, when the third is present the chord has the maximum rating ($m = 5$), so that naturally the third is to be included if possible.

It will be seen from the tabulation that the three inversions of the dominant 7th chord have a number of satisfactory forms. These are used frequently (P., pp. 103–109).

28. The Dominant 9th Chord

The dominant 9th chord has next to be considered. Since it involves five notes, all of them cannot be present in four-part harmony. However, the root, G, and the dissonant ninth, A, are present of course.

It is to be pointed out first of all that the requirements of regularity which have been imposed are in accordance with the usual rules and also with the theory elaborated above. To this end we may quote (P., pp. 161, 164): "Inversions of the chord of the ninth are . . . very rare." "As the major ninth if placed below the third of the chord will be a major second below that note, it will frequently sound harsh in that position. It is therefore generally better to put the ninth above the third." Furthermore we observe that for an irregular dominant 9th chord, Cd is negative, I is at most 2, and D is at most 0, so that no irregular dominant 9th chord can rate as even passable. Consequently we need only consider the regular dominant 9th chords.

It is easy to determine the only passable forms of the incomplete dominant 9th chord, in which the characteristic seventh, F, is absent. Evidently, since the dissonant ninth falls above the root, Cd is at most -1. Also since F is absent, the interval of the diminished fifth is not present, and I is at most 1. Hence I must be 1 and D must be 2, if the chord is to be passable. This means that the root G must be in the bass and dominating, and that the third B is also present. Hence the only passable possibilities are the following:

$$\begin{array}{cc} \text{G–B–D–A} & \text{G–B–A} \\ 2 & 2 \end{array}$$

Only the complete forms of the regular dominant 9th chord remain to be considered. The tabulation of these is as follows:

$$\begin{array}{cccc} & & {}^{*}\;\;{}^{*} & {}^{*} \\ \text{G–B–F–A} & \text{G–D–F–A} & \text{G–F–A} & \text{B–F–G–A} \\ 4 & 2 & 2-2-2 & 0 \\ {}^{*} & & {}^{*} & {}^{*} \quad\;\; {}^{*}\;{}^{*}\;{}^{*} \\ \text{D–F–G–A} & & \text{F–G–B–A} \;\; \text{F–G–D–A} \;\; \text{F–G–A} \\ 1 & & 1 \qquad\quad -1 \qquad\; -3\;0\;-3 \end{array}$$

Here the root position with the first inversion, and the second and third inversions appear in successive lines.

THE DIATONIC CHORDS

The statement to be considered in the light of all these results is the following (P., p. 161): "The seventh is almost always either present in the chord or, if not, it is added when the ninth is resolved. ... In the root position it is generally the fifth that is omitted; but if the root be not in the bass, it is seldom present at all. Inversions of the chord of the ninth are therefore very rare. ..." Evidently this statement corroborates the tabulation, which indicates that the only *good* form of the chord ($m = 4$) in four-part harmony is the fundamental position with the fifth absent.

29. The Dominant 11th Chord

Let us proceed similarly with the dominant 11th chord. It is again found that there are no irregular cases which are rated even as passable. Also if the chord is regular but incomplete, the possibilities in fundamental position are seen to be rated as follows:

$$\begin{array}{cccc}
 & & & * \\
\text{G–B–D–C} & \text{G–B–A–C} & \text{G–D–A–C} & \\
2 & 2 & 1 & \\
* \; * & * \; * & * \; * \; * & \\
\text{G–B–C} & \text{G–D–C} & \text{G–A–C} & \\
2-2-2 & 2-2-2 & 1-3-3 &
\end{array}$$

No inversions need be considered since they cannot rate higher than 0 in the most favorable case.

Thus it remains only to tabulate the complete forms of the regular dominant 11th chord with root, seventh, and eleventh present:

$$\begin{array}{cccc}
 & & & * \; * \\
\text{G–B–F–C} & \text{G–D–F–C} & \text{G–F–A–C} & \text{G–F–C} \\
4 & 3 & 2 & 3-1-1 \\
* & & & \\
\text{B–F–G–C} & & & \\
0 & & & \\
* & & & \\
\text{D–F–G–C} & & & \\
1 & & & \\
* & * & * & * \; * \; * \\
\text{F–G–B–C} & \text{F–G–D–C} & \text{F–G–A–C} & \text{F–G–C} \\
1 & 0 & -1 & -2 \;\; 1 \; -2
\end{array}$$

Here the fundamental position, the first, second, and third inversions appear in successive lines; the fourth and fifth inversions are not possible in

AESTHETIC MEASURE

the regular case. In the last of the third inversions, C is taken as the 'fourth above the bass,' since B is the expected third.

These results must be compared with the following statements: "As the eleventh is a dissonance, the usual resolution of which is by descent of a second [i.e. descent to the *third*], the third is mostly omitted in accordance with the general principle [of resolution]. . . . Either the fifth or ninth of the chord is also generally omitted; but the seventh is usually present . . ." (P., p. 172). "Owing to the harsh dissonance of the eleventh against the third, the first inversion of this chord is very rare" (P., p. 174). "The second inversion of the chord is much more common than the first" (P., p. 174). "The third inversion of the chord is so rare that we are unable to give an example of it" (P., p. 175). "The fourth and fifth inversions are also very seldom met with" (P., p. 175).

In making such a comparison let us recall the fundamental fact that the aesthetic measure proposed is a measure of the *single chord* taken in a definite tonality. Hence, according to the theory, the first complete chord listed, G–B–F–C, is the best of all the dominant 11th chords, when taken singly. In my judgment this is true. It seems to me also to be true that, of the other forms, the fundamental positions rated as passable are precisely the ones which are tolerable to the ear, and that the second and third inversions which have highest rating ($m = 1$) among the others come next in order.

If we accept this as the fact, the following explanation of the first statement quoted seems to be the correct one: It is not possible to use this best form, as it stands, in a sequence of chords, for, as stated by Prout in the first quotation, there would be a difficulty in the usual resolution to the dominant chord; in fact the thirteenth, C, must fall to the third, B, which cannot then be present in the chord. However, if the third is omitted from the chord, while the eleventh drops to the third in the resolution, the third may be regarded as *present by anticipation*.

With this interpretation, it is also reasonable to think of the second inversion D–F–G–C as D–F–G–(B–)–C with $m = 2$. Here we count the interval value I as though B were actually present, but do not admit the semitone dissonance B–C as actual. Hence the comparative commonness of the second inversion seems to be what might be expected.

THE DIATONIC CHORDS

30. The Dominant 13th Chord

In dealing with the dominant 13th chords, we are immediately reduced to those of regular type, as the only ones which can be passable, just as in the case of the dominant 9th and 11th chords. Of the incomplete chords, which lack the seventh, it is again clear that none can be passable which are not in fundamental position; and that even then the root must be dominating. Thus the only passable incomplete forms are found to be the following:

$$
\begin{array}{cccc}
\text{G–B–D–E} & \text{G–B–A–E} & \text{G–B–C–E} & \text{G–D–E} \\
3 & 2 & 2 & 2
\end{array}
$$

In the complete case, when G, F, E are all present in the chord, the following possibilities arise:

$$
\begin{array}{ccccc}
\text{G–B–F–E} & \text{G–D–F–E} & \text{G–F–A–E} & \text{G–F–C–E} & \overset{*\ *}{\text{G–F–E}} \\
4 & 2 & 2 & 2 & 2\ -2\ -2
\end{array}
$$

$$
\overset{*}{\text{B–F–G–E}} \\
0
$$

$$
\overset{*}{\text{D–F–G–E}} \\
0
$$

$$
\begin{array}{ccccc}
\overset{*}{\text{F–G–B–E}} & \overset{*}{\text{F–G–D–E}} & \overset{*}{\text{F–G–A–E}} & \overset{*}{\text{F–G–C–E}} & \overset{*\ *\ *}{\text{F–G–E}} \\
1 & -1 & -1 & -1 & -3\ 0\ -3
\end{array}
$$

Here the fundamental positions and the three possible successive inversions are given in the successive lines.

Concerning the dominant 13th chord, Prout says (P., p. 186): "Though an enormous number of combinations . . . are *possible*, comparatively few are in actual use . . . ." He lists the three principal types as follows: "I. Root, third, and thirteenth; II. Root, third, seventh, and thirteenth; III. Root, third, fifth, and thirteenth."

Now the first of these chords is merely the mediant chord, although it may function as a derivative in a sequence of chords. This chord need not occupy us further since it has already been rated as passable in itself ($m = 2$). The second form is "the commonest and most useful form of the chord" (P., p. 188), and is the only one rated above as good ($m = 4$).

The third form is 'much rarer than the preceding' (P., p. 190), and is the only one rated as fair ($m = 3$).

31. The Derivative Chords

On the same basis we can at once determine all usable forms of derivative chords.*

$$V_7$$

The only possible derivative here is the dissonant diminished fifth B–D–F already considered (sections 23–26).

$$V_9$$

The only possible derivative here is made up of some or all of the notes of B–D–F–A, the so-called 'leading 7th chord.' The notes B and A must actually occur, with A above B.

An examination of the possibilities yields only the following passable cases:

B–D–F–A   D–F–B–A   F–B–D–A
2   2   2

Of the three examples given by Prout of the use of this chord in a major key (P., pp. 167–168), the first, taken from Graun's *Te Deum*, illustrates the 'root position' of the leading 7th chord, with doubled leading note. This chord is extremely harsh as one would expect; its rating in this form is 0. The second example, taken from Mendelssohn's *Variations*, Op. 82, illustrates the first inversion in four-part form, rated above as passable. The third example, taken from his *St. Paul*, illustrates the second inversion above, rated as passable.

As Prout notes (P., p. 169), it is the leading 7th chord in a minor key that is of first importance; this is the chord of the diminished 7th (section 16), "the chief derivative of the chord of the ninth, and its most frequently used form." Indeed this chord is useful also in major harmony for chromatic purposes and for modulation.

Since we are confining attention strictly to diatonic chords, the diminished 7th chord does not enter into consideration.

* In the tabulation such derivative chords are considered to have major character of course.

THE DIATONIC CHORDS

V11

The only possibilities in question here are made up of a selection of the notes B–D–F–A–C, in which C is present. If B is present also, there is the harsh discord of a major seventh. In this case if F is not present the chord is incomplete (as a dominant chord) so that Cd is -2 and I is at most 1; moreover F and G are not present and the primary note C is dissonant, so that D is at most 0. Here then m cannot exceed -1 and the chord is not usable even in sequence according to the theory.

When F is present, however, Cd is at most -1, I is at most 2, and D is at most 0 since F is not consonant (section 21). Hence m cannot exceed 1. It is therefore easy to understand why "the derivative of the first inversion [of the dominant 11th chord] is as rare as the inversion itself, and for the same reason — the harshness of its dissonance" (P., p. 175).

Thus there remains to consider only the notes of the 'supertonic 7th chord' D–F–A–C, in which D as well as C must be present. Taken as a derivative and in complete form, this chord is not passable unless the primary note F is dominating and not doubled.* In four-part harmony there is one passable form, namely that of the first inversion of the supertonic 7th chord with all of its notes present. This explains why this first inversion is "one of the most frequently used of the derivatives of the chords of the eleventh" (P., p. 178).

V13

Here we have to make a selection of the notes B–D–F–A–C–E in which E is to be included.

If the characteristic seventh, F, is not present, Cd is at most -1 and I at most 1. Hence, in order that the derivative chord rate as usable even in sequence, the dominating note must be a consonant primary note. Since F and G are lacking, this must be C, which, according to our definition (section 15), is not possible in a true derivative chord.

If F is present, there is semitone dissonance, and Cd is -1. But here I may be as much as 2 if B is also present. Even in this case, however, the chord is at best usable in sequence ($m = 1$). If B is not present, the chord is not so usable.

* The chord is not passable since a doubled F functions as a doubled dissonant note relative to the suggested dominant G.

With this result are to be compared the statements (P., p. 191) that the derivatives of the dominant 13th chord in common use are "nearly all chromatic" and so not diatonic, and that "the only important diatonic derivative is that in which the root, third, fifth of the chord are all wanting."

It is to be observed, however, that because of the unquestionably harsh dissonance of this 'subdominant 7th chord,' it is usual to *prepare* the dissonant seventh or root, that is to prolong one of these notes from a consonant note of the preceding chord, and to *anticipate* the leading note. If we regard these devices as diminishing the index of dissonance from 2 to 1 and introducing the diminished fifth, the chord thereby becomes passable. Such preparation and anticipation occur in the single example in a major key given by Prout (P., p. 191).

In connection with all derivative chords, it should be noted that the aesthetic effect depends on the intensity and definiteness with which the lacking dominant note, G, and its third, B, if lacking, are suggested by the musical context. Hence, under certain circumstances, the derivatives may have an aesthetic effect hardly less satisfactory than that of the best corresponding dominant chords.

32. Irregular Chords

It is easily shown that no irregular diatonic chord can rate even as passable in its aesthetic measure m.

In fact the chord value Cd of such a chord is at most -2, because of its dissonance and irregularity, while the value D of the dominating note is at most 1. Hence $Cd + D$ is at most -1. Thus I must be at least 2 if the chord is to be usable in sequence; and so the notes B and F must be present. Furthermore a consonant primary note must occur if D is to be 1. But C is dissonant with B, and G is dissonant with F. Hence there is no irregular chord which can be regarded as usable even in sequence.

Such irregular chords fall under the head of 'secondary discords' which are not considered to be derivatives. According to Prout (P., p. 193) "much more importance was formerly attached [to these discords] than is the case at the present day."

THE DIATONIC CHORDS

In the first of the three examples given by Prout (P., pp. 194, 195), taken from Handel's *Joshua*, 'secondary 7th discords' are used in the sequence of a pattern and there is a rapidly moving soprano over slowly changing lower notes. Thus the lower notes function more or less as drone notes, and this steps outside of the restricted domain of isolated chords to which we are confining attention.

In the second example taken from Bach's *Fugue in E Minor* there is preparation of the secondary 7th discords, so that the chords are irregular only because of a sustained note of preparation continued from the preceding chord. This also lies outside of the domain under consideration.

The third example, taken from Cherubini's *Mass in F*, shows the "much rarer" secondary 9th discords used in sequence. The strong contrapuntal play between the bass and soprano parts justifies an unusually harsh sequence of dissonant chords.

33. Summary

In this chapter, then, we have recalled the origin and development of the Western diatonic scale and its various chords. The genetic account makes clear how the simple elements of order appearing in our formula for the aesthetic measure m of a diatonic chord arose.

These elements of order are usually regarded as so obvious that they are mentioned only incidentally in textbooks on harmony. According to our quantitative theory, the associative apparatus of the mind automatically weighs these elements, positive and negative, and the resultant tone of feeling determines the aesthetic value, measured by m. The diagram of section 22 may be regarded as a schematization of this process.

The rating of chords so obtained has been found to be in agreement with the usual empirical rules of classical harmony concerning the use of chords. But of course the theory goes much further than these rules since it affords the means of comparing any two diatonic chords whatever.

In connection with our theory of the single diatonic chord, it should not be forgotten that the musical quality of individual chords is often a matter of small importance relative to other valid musical effects.

CHAPTER VI

DIATONIC HARMONY

1. The Problem of Chordal Sequences

WHEN two chords of the diatonic scale, no matter how pleasing in themselves, follow one another without due regard to the progression of the various parts, the effect is unsatisfactory, and the sequence is unsuitable for musical purposes. For instance, if the four-part dominant chord in fundamental position, G–g–b–d′, be followed by the adjacent sub-dominant chord F-f-a-c′, the sequence obtained is disagreeable notwithstanding the fact that both of the chords are highly satisfactory. Thus the varying aesthetic quality of chordal sequences depends not only upon the aesthetic factors involved in the constituent chords but also on other factors involved in the method of transition.

The classification of such chordal sequences presents an important aesthetic problem, fundamental for harmony, which is far more complicated than that of the single chord. It is our aim to deal with this problem of chordal sequences from the point of view of aesthetic measure. For the sake of definiteness we shall confine attention to the important case of diatonic harmony in four distinct parts. The key will be taken as that of C major.

We shall employ an arrow to indicate the direction of the progression; thus Ia→Vb would indicate a sequence in which the passage was from the tonic chord in fundamental position to the dominant chord in its first inversion. Use will sometimes be made of the double arrow ⟵⟶, as for instance I⟵⟶V, to indicate that the sequence is to be taken in both orders.

2. The Method of Attack

In order to obtain an aesthetic formula for chordal sequences, we need first to obtain a suitable definition of the complexity, C. Now it has been pointed out that the complexity, C, of all single chords must be regarded as the same. Evidently the complexity of all sequences of two chords must also be taken to be the same. For definiteness we adopt the value 1 for C. With this simplification the basic formula, $M = O/C$, reduces to the simpler formula: $M = O$. In other words the problem is reduced to the determination of the elements of order, O, involved in such sequences.

We propose to attribute the total aesthetic effect to the individual effects of the chords together with the effect of the transition from the first chord to the second; thus, if m_1 and m_2 denote the aesthetic measures of the first and second chords respectively, as determined by the preceding theory, and if t denotes the similar aesthetic measure of the transition, we shall regard the aesthetic measure of the sequence, M, as given by the sum of these three components:

$$M = m_1 + t + m_2.$$

The fundamental problem before us is that of specifying the transition measure t.

3. Further Limitation of the Problem

It would be desirable to present a theory applicable to all chordal sequences whatsoever, or at least to all sequences in which the individual chords are regular. Nevertheless it is proposed here to consider only those 'regular chordal sequences' in which (1) the leaps in the individual voices are not excessive, and (2) the similarity in function of the voices is not too great. Of course it is in no way suggested that a slightly extended theory would not suffice to deal with all chordal sequences. Such a theory would include two further negative elements of order, corresponding to excessive leaps and similarity.

The arbitrary and fluidic character of the specific limitations which we state below must be strongly emphasized. In general, other things being equal, it is desirable that the leaps in the voices are not large and that the voices perform dissimilar functions. But, as soon as any specified limitation of this kind interferes with free musical expression, it should be ignored.

Nevertheless the great body of classical music usually observes such limitations.

4. The Limitation of Leaps

The urgency for some limitation as to leaps is obvious. If the leaps are excessive, not only are the voices likely to fall outside of their natural compasses, but an unusual effort is required to make the leap, and this affects the hearer unpleasantly. The step of a second or the leap of a third up or down is more easily made in general than a leap of a larger interval. Scarcely ever is an interval of as much as an octave used, and this ordinarily occurs in the bass. The rules to be stated presently are such as to take account of these and similar facts, and thus to insure reasonable ease of execution in singing the parts.

Inasmuch as the bass part frequently occurs as the root of a primary chord, and the progressions between roots of primary chords are often by perfect fourths or fifths, up or down, the bass voice becomes used to these leaps. In fact such a 'harmonic step' in the bass voice is made almost as easily as a 'melodic step' in the upper voices. The leap of an octave is also made frequently in the bass part in order to avoid a stationary note.

In the upper parts a leap of as much as an octave usually occurs only when the second chord is another position of the first; for in this event there is greater ease of execution of the leap than there is when the chord changes. Only one leap of a fourth or more is agreeable in the upper parts when the chord changes, however, and if the bass leaps by as much as a sixth, no leap in the upper parts should be more than a third.

In general, no part should leap by a dissonant interval. An obvious reason for this feeling is that the dissonant interval is unpleasantly suggested; a further and perhaps more important reason is that the consonant intervals are the ones of natural harmonic origin and are therefore more easily sung.

There are, however, two exceptions. In the first place the interval of the diminished fifth (B up to F, or F down to B), which is only slightly dissonant and which is a pleasing interval, is tolerated. However, the equivalent augmented fourth (B down to F, or F up to B) is not allowed. The distinction felt may be explained as follows: both notes involved are harmonically related to one another because they are essentially

DIATONIC HARMONY

overtones of the dominant; however, as an overtone of the dominant, the leading note is lower in position than the subdominant, and in consequence the interval is only tolerated when the notes occur in this natural order.

In the second place the leap of a minor seventh, from the dominant up to the subdominant or *vice versa* is permitted for a similar reason; this dissonant interval lies in the constantly used dominant 7th chord. However, such a leap may occur only in the extreme parts, since in the middle parts it leads to forbidden crossing of the voices.

The reason that such crossing or overlapping of the parts is avoided is that any voice has a tendency to stop at a note just taken by another voice rather than to proceed beyond it. The effort required to resist this natural tendency is distinctly noticeable.

Such are the usual limitations governing the leaps in the parts, and we shall embody them in the following three simple rules:

No part is to leap across a note just sounded in another part. When the chord does not change, there is no further restriction in the motion of the parts.

Otherwise, at most one leap of a fourth or more is allowed in the upper parts, and this is to be less than an octave. No leap of more than a twelfth (octave plus fifth) is allowed in the bass. If the leap in the bass is as much as a sixth, no leap in the upper parts exceeds a third.

All leaps are to be by consonant intervals, with the possible exception of a leap of a diminished fifth, or of a minor seventh from dominant up to subdominant or *vice versa*.

One other requirement may be mentioned although, strictly speaking, it does not apply to a sequence of only two chords: if a part leaps by as much as a sixth in one direction, it is desirable for the part to do so with reversal of the direction of motion; obviously this reversal of direction relieves the unusual vocal effort involved.

Such arbitrary rules of limitation as those imposed in this section and the next are often successfully broken. But, inasmuch as we are confining attention to music *with a definite allowance of means*, it is permissible to impose requirements which are generally observed in the large body of classical music.

5. Limitation of Similarity of Function

The basic reason for some limitation of the similarity of function in two voices lies in the requirement of a homogeneous use of the voices in musical composition: either the voices should be similarly related throughout, as by successive octaves, or not so related anywhere in a musical composition. Thus distaste for certain types of similarity has taken definite form.

The similarity between two voices varies with the degree of harmonic relationship between the notes involved and is enhanced by 'similar motion,' when both voices rise or fall. On the other hand, any similarity is much less marked in the case of contrary motion, and is felt to be less objectionable between the strong primary chords and in the inconspicuous middle voices. Such are some of the general considerations which have operated to determine ordinary practice.

The types of similarity to which objection is felt refer particularly to 'consecutive' octaves, perfect fifths and perfect fourths, and to 'hidden' octaves and perfect fifths in which two voices move by similar motion to an octave or a perfect fifth.

The entirely explicit rules of Prout (P., pp. 25–33) are not formulated so that any associative basis for their curious structure is obvious. A nearly equivalent formulation, which displays the basic associative structure, is as follows:

Consecutive octaves and consecutive perfect fifths are forbidden in the case of similar motion (when there is almost complete identity of function). In the case of consecutive octaves they are allowed * by contrary motion between the primary chords with doubled primary notes (the best chords of all). In the case of consecutive perfect fifths it is required further that the extreme parts be not involved (because of the greater conspicuousness of the perfect fifths in these parts).

Consecutive perfect fourths are only forbidden by similar motion when one part is in the (prominent) bass part (for the third above the bass and *not* the fourth is desired).

Hidden octaves are forbidden when the voices move from a dissonant

---

* Providing that the sequence is satisfactory *otherwise*.

DIATONIC HARMONY

seventh or ninth to the octave (since the prominent dissonance is felt not to be properly resolved). Otherwise hidden octaves and hidden perfect fifths are forbidden only when in the (conspicuous) extreme parts; however, in case the second chord is a primary chord with doubled primary note (one of the best chords) they are tolerable as follows: in the case of hidden octaves, when both chords are primary chords in fundamental position and the upper part moves by step only (thus diminishing the similarity of function since the bass leaps); and when the second primary chord is not in fundamental position (so that the octave in question does not involve the conspicuous root and bass of the chord); in the case of hidden perfect fifths, when the first chord is also primary and the upper part moves by step (thus diminishing the similarity of function since the bass leaps); and when both chords form a dominant sequence (and so lie practically in the same chord).*

We shall not attempt to show that these general implicit limitations are co-extensive with Prout's entirely explicit rules.

6. The Law of Resolution

Before proceeding to enumerate the elements of order involved in chordal sequences, it is necessary to recall one further musical phenomenon, namely that of resolution of dissonance.

Music involving only consonant chords is lacking in interest to the ear. In fact dissonance, followed appropriately by consonance, is a most important method of obtaining contrast. Now in classical music the means by which such resolution of dissonance may be effected have been rather clearly defined. Let us consider a few very simple instances.

The notes of the dissonant interval B–f are seen each to be only a semitone distant from those of the consonant interval c–e. If the interval B–f be followed by the consonant interval c–e, the ear is satisfied and B–f is said to be 'resolved' on c–e. Similarly, when the dominant 7th chord G–B–f goes to G–c–e, with B–f moving to c–e as before while G is stationary, V_7 is resolved on I. Or if we take the interval of the minor seventh G–f of V_7, the dissonance is resolved on I by allowing the disso-

* The only case of this sort is iia → Va with the third of iia in the soprano.

nant note, f, to move to the nearest note of the tonic chord, e, while the note G is held.

Evidently in these examples, the dissonant notes of the chord move to the *nearest* positions in a consonant chord.

By comparison of various cases of resolution, these are found to obey a general 'law of resolution,' which may be formulated in the following way:

A dissonant chord goes to a consonant chord as follows: (1) a dissonant note of the first chord not found in the second chord goes to a nearest note of the second chord, which must not be a dissonant note of the first chord; (2) if the particular note thus designated has been taken by the bass part, or by an adjacent part, or by an octave of the given note, the note is free to move a step in the opposite direction, except for the tonic which may leap a third in the opposite direction; (3) the second chord must be the tonic, subdominant, or submediant chord via in case the first chord is V_7 or its derivative viib; in all other cases it must be the tonic, dominant, or submediant chord via.

In connection with the requirement (3) we observe that the chord V_7 and its derivative viib are too closely related to the dominant chord V to resolve satisfactorily upon V. Furthermore, resolution upon the submediant via is permitted because this chord is felt to be a kind of substitute for the tonic chord; in fact via has two notes in common with the tonic chord and is indeed the fundamental position of the tonic chord in the 'related minor' key of A.

The manner of handling dissonance is not always to pass directly from a dissonant chord to a consonant chord. In fact sequences of dissonant dominant chords and of their derivatives are often employed in the following order:

$$V_{13} \to V_{11} \to V_9 \to V_7, V, \text{viib} \to I \text{ or via}.$$

Here some of the chords may be omitted, or the ending may be of the form $V_7 \to IV \to I$. This or any other logical sequence of dissonant dominant chords may be employed in a musical composition. In such a case the resolution is of course effected at the moment of passing to a consonant chord by the method prescribed in the rule.

DIATONIC HARMONY

7. The Element of Resolution: $R = 4, -4$

We are now able to formulate briefly the various elements of order which (in our opinion) determine the aesthetic measure of transition, t.

Let us begin with the element of resolution, R, which is present when a dissonant chord proceeds to a consonant chord in accordance with the law of resolution of the preceding section.

We shall take R to have the index 4 if resolution is properly effected; if it is not so effected in a sequence of two such chords we shall take R to be -4, since the effect is very unsatisfactory. In any chordal sequence not of this type R is 0 of course.

The index 4 seems to be that which rates this element suitably in connection with the indices used in the aesthetic measure of the single chord. This means that the element R in the transitional effect is rated as about equally important with that of the effect of a good chord ($m = 4$ or 5).

8. The Cadential Element: $Cl = 4, -2$

Cadence may be regarded as a very mild form of resolution. For example, if we pass from the dissonant dominant 7th chord to the tonic, $V_7 \to I$, there is definite resolution. If we pass from the closely related consonant chord V to the tonic, $V \to I$, a similar cadential effect is felt. The two fundamental forms of cadence are the above so-called 'authentic cadence' $V \to I$ and the less frequent 'plagal cadence' $IV \to I$. The authentic form is usually found at the end of a musical composition; the plagal form is also employed, but much less frequently, as in the familiar *Amen*.

By a 'cadential sequence' we shall mean, however, not only sequences of these types $V \to I$, $IV \to I$ but also certain analogous sequences, which we proceed to enumerate.

In the first place, the submediant in fundamental position, via, may be regarded as a kind of substitute for the tonic chord (see the preceding section 6). Thus the sequence $V \to$ via is the frequently used 'interrupted cadence' and has a definite cadential effect. Similarly the chord of the leading note, viib, functions in many instances as V_7 (V) so that the sequences viib $\to$ I, via are felt to be cadential. In the case viib $\to$ via, however, when neither chord is primary, the leading note of viib must go to the tonic ($7 \to 1$) for a satisfactory cadential effect.

The cases so far enumerated have the authentic cadence, $V \to I$, for their prototype. There is one more of this general character, namely, the sequence from mediant to submediant, iii $\to$ vi, for this functions as the authentic cadence in the related minor key.

In the case of the less important plagal cadence, the only analogous form felt to have cadential quality is the corresponding one in the related minor key, ii $\to$ vi.

We shall regard all of these types of sequences as cadential, and the corresponding cadential element of order, Cl, will be given the index 4, the same as the index for the element of resolution, R.

Moreover we shall regard the same types of sequences, *taken in the reverse order*, as cadential, so that Cl is 4 in these cases also.

The question may be asked as to why the inverted order is allowed here, although these are not ordinarily considered to be cadential. This may be explained in part as follows: The plagal cadence is equivalent to a reversed form of the authentic cadence as far as the intervals involved are concerned; this reversibility extends by association to the other forms.

It remains to refer to two other cases when an unpleasant or *false* cadential effect is produced. In the first case, the leading tone goes to the tonic ($7 \to 1$) in an extreme part (as usually happens in the fundamental direct form of authentic cadence $V \to I$) and yet there is no cadence of the above named types; under these circumstances there is a feeling of disappointed expectation. In the second case, IV in fundamental position goes to the related tonic minor vi in fundamental position; thereby a plagal cadence IV $\to$ I is suggested, but the *two* common notes of IV and vi negate the cadential effect so strongly that the effect is again one of disappointed expectation.

Thus the various types of positive cadential sequences are taken to be as here presented:

$$V, viib \longleftrightarrow I, via \ (7 \longleftrightarrow 1 \text{ in viib} \longleftrightarrow via); \qquad iii \longleftrightarrow vi;$$
$$IV \longleftrightarrow I; \qquad\qquad\qquad\qquad\qquad ii \longleftrightarrow vi.$$

For these, Cl is 4.

DIATONIC HARMONY

The types of false or negative cadential sequences are the following:

7 → 1 in an extreme part but without cadence; IVa → via.

For these, Cl is -2.

In any other case Cl is taken to be 0.

9. The Element of Dominant Sequence: $D = 4$

In the classification of dominant chords V, V7, V9, V11, V13, it has been seen in the preceding chapter that certain forms of these are to be regarded as satisfactory. When a sequence of two of these is heard, they are pleasantly united by a common dominant quality whose importance is obvious.

Moreover, certain consonant chords besides V may be imbued with dominant quality. The principal consonant chords of this kind are: viib, which may be regarded as a derivative of V7; ii, as a derivative of V9; and iii, as a derivative of V13. It is to be noted that the first inversion of iii has the dominant in the bass, and in its only passable form has three of its four notes in the dominant chord.

However, in the sequences V7, viib → V there is little or no sense of dominant *sequence* since viib appears naturally as part of the slightly dissonant V7, which includes V and calls for resolution on the tonic chord. On the other hand, V → V7, viib form dominant sequences, since V7 and viib extend beyond V.

Furthermore, whenever the chord viib comes first, its weak dominant character is not felt when it is followed by ii or iii. Likewise a sequence ii ⟷ iiia is not felt to have dominant quality, since ii is not strongly dominant and iiia is scarcely recognized as dominant. Under such circumstances also, the element of dominant sequence, D, is taken to be absent.

In case ii comes first in ii ⟷ V, viib, the submediant, 6, must proceed to the adjacent 5 or 7, as in the resolution of V9. Likewise if iii comes first in iii ⟷ V, viib, the mediant 3 must proceed to the nearest available note 4 or 2; and in case iii is in its weak dominant position iiia, the note 3 must be in the soprano (as in the case of V13) and must proceed to the nearest note as in a resolution of V13. If these additional conditions are not satisfied, D is taken to be absent.

Finally, if resolution occurs, D is not counted independently of R.

Hence, subject to the further conditions just stated, we regard the element D of dominant sequence as present in any sequence of two chords ii, iii, viib, V, V7, V9, V11, V13, excepting V7 → V and viib → ii, iii, V and ii ⟷ iiia. Under these circumstances D is taken as 4, and otherwise as 0.

10. The Elements $SF = 4$ and $RF = 4$

In passing from one three-part consonant chord in closest possible position to another, there are certain special cases in which the transition is unusually smooth. These are those in which certain undesirable contingencies are avoided, namely (1) stationary bass or soprano, (2) forbidden crossing of the parts, (3) forbidden consecutive perfect fifths, (4) second inversions, which are only passable in the primary chords.

An examination of the various possibilities is easily made. If, for example, we designate the root, third, and fifth of the first chord as 1, 3, and 5, we find only the following five general possibilities:

$$3\text{-}5\text{-}1 \to 4\text{-}6\text{-}2,\ 5\text{-}7\text{-}3,\ 2\text{-}4\text{-}7,\ 1\text{-}3\text{-}6;\ 1\text{-}3\text{-}5 \to 2\text{-}4\text{-}7.$$

Hence such 'close harmony' arises in the case of successive first inversions in which the root falls or rises by one or two steps (SF), and also when the first chord in root position is followed by a first inversion with root falling one step (RF). In the first case we say that the element SF is present, and in the second case that the element RF is present. Both of these elements are given an index 4.

It will be found that the usual rules are consistent with the assignment of definite aesthetic value to these special types SF and RF of root progressions.

The smoothness of a chordal progression SF may be verified by playing an arbitrary set of first inversions in which the bass steps or leaps at most two steps as one desires. The type RF is equally smooth but does not admit of incorporation in a progression of this kind because there is no means of returning to fundamental position, once it has been left.

11. The Element of Progression: $P = 2, 0, -2$.

If, in passing from the first chord of a sequence to the second, it is possible to 'borrow' a note of the first chord as a consonant part of the

DIATONIC HARMONY

second, the transition from one to the other is felt not to be unnatural. Since V is so closely associated with viib, through V7, it is also permitted to borrow the dominant note for use in viib.

If two or more notes can be so borrowed, there is felt to be a sense of positive harmonic progression in passing from the first chord to the second: for, the anticipation provoked by the first chord is then fulfilled in a suitable manner.

Moreover, an effect of melodic progression is obtained when the two chords have no notes in common, provided that each note of the first chord proceeds to a nearest note of the second chord, except in so far as the bass or an adjacent part or an octave of this note has proceeded to such a nearest note, in which case the note is free to move a step in the opposite direction. Furthermore the tonic may leap a third in the opposite direction, and a note freed by its octave may move at pleasure in the opposite direction.

In all of these cases the 'element of progression' P is taken to be 2. Otherwise, if there is resolution, or if the sequence is dominant, or if one and only one note can be borrowed, P is taken to be 0. In every other case the two chords appear as unrelated and P is taken to be -2.

In illustration of progression value, P, we may consider the following sequence of chords I → V → ii → vi → iii → vii → IV → I.* This sequence

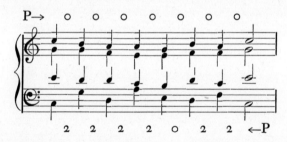

lacks harmonic progression so that $P = 0$, whereas in the reverse order, such progression exists.

The simple positive elements R, Cl, D, SF, RF, and P are all of unquestionable aesthetic importance. These are the only positive elements which we shall take account of in our theory. The remaining elements are all negative in their aesthetic effect.

* See D. N. Tweedy, *Manual of Harmonic Technic*, Boston (1928), p. 106.

12. The Negative Element $FR = 4$

If the first chord is a first inversion and the second chord is in fundamental position while the root falls one step, the effect is in general unsatisfactory.

In fact this is the only type of root progression which, if executed in 'close harmony' by three voices, requires two forbidden crossings of the voices, gives rise to chords without common note of connection, and furthermore does not start from a strong root position.

For such a sequence we take the element FR to be 4; in any other case we take FR to be 0.

This negative element is in part recognized by Prout in his rule VI of root progression:* "As a general rule, whenever the root falls a second, *the second of the two chords should be in its first inversion.*" In other words, the progression is to be of the type SF or RF, and not of the type FR or SR.† This statement is clearly in agreement with our statements above: for, both the elements SF and RF have been taken to be positive, while FR has been taken to be negative, and SR is impossible in close harmony on account of forbidden consecutive fifths.

13. The Negative Element of the Mediant: $Mt = 2, 4$

Among the fundamental triadic chords, the mediant chord is generally felt to occupy the least favorable position. For, among these chords, the dominant and subdominant chords, being of the same constitution as the central tonic chord are felt to be 'primary'; furthermore, the leading note chord and the supertonic chord are dominant enough in their quality to be felt to be related to the primary dominant chord; and of the two remaining chords (the mediant and submediant), the submediant fulfills the important function of substitute for the tonic chord itself. Thus the mediant chord, of doubtful dominant quality in its fundamental position and having no special function, is felt to be of least value.

For some such reason the presence of the mediant chord in a chordal sequence produces a well known unfavorable effect.

---

* *Counterpoint: Strict and Free* (third edition), p. 30.
† That is, a sequence of two fundamental positions with root falling a second.

DIATONIC HARMONY

On account of this fact the negative 'element of the mediant' Mt with index 2 will be generally attached to any chordal sequence in which the mediant chord is present; in certain exceptional cases, specified below, the index 4 will be assigned.

Furthermore, in case iii is not present but the characteristic first and third of vi and iii appear in the first and second chords respectively, while the bass leaps either from the submediant 6 or to the mediant 3, or does both, all as in a characteristic cadential sequence from vi to iii, we also take Mt to have the index 2.

The reason for this is obvious: under these circumstances the mediant chord is strongly suggested even though it is not actually present.

However, if iii is present and at the same time the preceding conditions are met, we take Mt to have the index 4, since the chord iii is then still more strongly suggested.

Finally, if there is a passage from the mediant chord in its fundamental position to the distant tonic chord in fundamental position, there is a strong effect of disjunction, so that Mt is also assigned the index 4 in this case. The inverse passage Ia → iiia, starting from the firm fundamental position of the tonic chord, does not produce a similar unfavorable effect.

14. THE NEGATIVE ELEMENT OF THE LEADING NOTE: $LN = 2, 4$

As its name indicates, the leading note has a distinct tendency to move to the adjacent tonic note, and, in a lesser degree, to the adjacent submediant, which is the tonic of the related minor chord. Hence, if the leading note appears in the first chord but not in the second, and does not step to an adjacent note, there is a definite feeling that the leading note has not been properly dealt with.

Moreover, whenever the leading note leaps more than a third, or is transferred from any part of the first chord to the bass of the second chord, the same feeling arises.

Again, in case the leading note is present in both chords, thus strongly suggesting a dominant sequence (note that iii, V, and viib are dominant), while the sequence is not dominant in point of fact, the leading note is felt to be particularly out of place.

Thus we assign to the negative element of the leading note, LN, an index 2 in these cases, excepting in the last case when we assign to it an index 4.

15. The Negative Element of Stationary Notes: $SN = 2, 4$

When an extreme part is held in a sequence of two chords, the effect is monotonous. Even if the bass rises or falls an octave, the effect remains much the same. A similar displeasing effect is felt when two inner voices are held, or when one inner voice is held and the leading note is left by one voice and taken again by the adjacent voice. Furthermore, if two of the voices are interchanged in position, as C and E going to E and C respectively, a similar unfavorable effect is produced.

In these cases we shall say that the negative element of stationary notes, SN, is present, and give SN an index 2 or 4, according as two parts or more than two parts are involved.

However, we do not allow more than two voices to be actually stationary, and exclude the case of both stationary bass and soprano voices, since chordal sequences are unsatisfactory under these circumstances.

16. The Negative Element of Dissonant Leap: $DL = 2$

The dissonant leap of a diminished fifth, or of a minor seventh to or from the dominant, although allowable (section 4), is not in itself agreeable, since the voice moves more naturally by a consonant interval than by a dissonant one.

In the case of such a dissonant leap we introduce a corresponding negative element of order, DL, of index 2.

17. The Negative Element of Similar Motion: $SM = 2$

If all the parts move up or if all move down, there is an undesirable effect of similarity of function produced by the similar motion, to which corresponds the negative element SM with index 2.

It is usually possible to arrange chordal sequences of a given type so that not all the parts move in the same direction.

18. The Negative Element of Bass Leap: $BL = 4$

If the perfect fourth above the bass is a note of a chord, the chord is in its second inversion and is felt not to be in normal position, since in

DIATONIC HARMONY

normal position the third and not the fourth above the bass is found in the chord. In consequence the bass voice moves readily to the bass in such a chord only by step, or by leap from a fundamental position; and it leaves such a position readily only by step.

Hence, if a chord of this kind is approached or left by leap, but not from a fundamental position, the negative element *BL* is taken to be present, and is assigned the index 4.

This completes our list of negative elements of order: *FR, Mt, LN, SN, DL, SM, BL*. All of these deal with negative aesthetic factors whose importance is scarcely questionable.

19. Recapitulation of Definition of *M* for Chordal Sequences

Let us recapitulate: The aesthetic measure *M* of a regular sequence of two chords is defined as the sum of the aesthetic measures of the two constituent chords m_1 and m_2, and of the transition value t:

$$M = m_1 + m_2 + t.$$

The transition value t is the sum of the following elements:

(1) *The Element of Resolution: R = 4, −4*

The element *R* is 4 when a dissonant first chord resolves on a consonant second chord, in accordance with the rule of section 6. If this resolution is not effected, *R* is −4.

(2) *The Cadential Element: Cl = 4, −2*

The element *Cl* is 4 if the sequence is of one of the following types:

V, viib ⟷ I, via (7 ⟷ 1 in viib ⟷ via);   iii ⟷ vi;
IV ⟷ I;   ii ⟷ vi.

The element *Cl* is −2 in the following types of sequences:

7 → 1 in extreme part but without cadence; IVa → via. Otherwise *Cl* is 0.

(3) *The Element of Dominant Sequence: D = 4*

The sequence involves two of the following chords ii, iii, viib, V, V_7, V_9, V_{11}, V_{13}, with V_7 → V and viib → ii, iii, V and ii ⟷ iiia excepted.

In the non-excepted cases D is taken as 4, provided the following further requirements are satisfied: in sequences ii → viib, V and iii → viib, V, the submediant in ii and the mediant in iii move by step; in iiia → V the mediant must be in the highest part of iiia.

For all other cases D is taken to be 0.

This element 0 is not counted if R is counted.

(4) *The Elements SF and RF:* $SF = 4$, $RF = 4$

These elements SF and RF are present respectively when there is a sequence of two first inversions in which the bass moves up or down one or two steps, and when a fundamental position is followed by a first inversion, with the root falling one step.

(5) *The Element of Progression:* $P = 2, 0, -2$

The element P is 2 if two notes of the first chord can be borrowed as consonant notes for the second chord. The borrowing of a dominant note for use in viib is permitted. The element P is also 2 when no notes of the first chord can be so borrowed, provided that each note of the first chord goes to a nearest note of the second chord, except in so far as the bass part or an adjacent part or an octave of this note has proceeded to such a note, in which case the first note is free to move a step in the opposite direction; moreover, the tonic is allowed to move a third, and the motion of a freed octave in the opposite direction is unrestricted.

Otherwise, the element P is 0 if one consonant note only can be so borrowed, or if there is resolution, or if the sequence is dominant; P is -2 in every other case.

(6) *The Negative Element:* $FR = 4$

The sequence involves a first inversion followed by a chord in fundamental position, with the root falling one step.

(7) *The Negative Element of the Mediant Chord:* $Mt = 2, 4$

The element Mt is 2 if iii is present but the sequence is not of one of the two types: iiia → Ia or (6, 1, *) → (3, 5, *) with bass leap from 6 or to 3 or both; also, if iii is not present but the sequence is of the second type; Mt is 4 if iii is present and the sequence is of one of these types.

DIATONIC HARMONY

(8) *The Negative Element of the Leading Note: LN = 2, 4*

If the leading note leaps and is not found in the second chord, or leaps more than a third, or is transferred to the bass, LN is 2. If the leading note occurs in both chords of a non-dominant sequence, LN is 4.

(9) *The Negative Element of Stationary Notes: SN = 2, 4*

The element SN is 2 if an extreme part or two parts are held stationary (inclusive of a leading note taken and left by adjacent voices) or interchanged. If three parts are involved in this way, SN is 4.

(10) *The Negative Element of Dissonant Leap: DL = 2*

One or more of the parts leap an allowed dissonant interval (section 4).

(11) *The Negative Element of Similar Motion: SM = 2*

All of the parts move up or down, but not in the same chord.

(12) *The Negative Element of Bass Leap: BL = 4*

The sequence contains a second inversion which is approached or left by leap, but not from a fundamental position of the first chord.

20. Comparison with Prout's Classification

Fortunately the aesthetic classification of the principal types of chordal sequences has been effected empirically; for instance, in Prout's *Counterpoint; Strict and Free* (third edition), p. 32, a table of sequences involving the fundamental positions and first inversions of the triadic chords, is given. These are rated by Prout as 'good,' 'possible,' and 'bad.'

It is essential of course that the results of our theory be in fair agreement with these tabulated results which are the results of observation and experience. Indeed I have used his table as a basic aid in the aesthetic analysis of chordal sequences.

In order to effect the comparison of Prout's table and these results, the best positions of all the possible 144 types of chordal sequences involving the fundamental positions and first inversions of the triadic chords have been determined on the basis of the definition of aesthetic measure given above. The aesthetic measures are entered in the table

AESTHETIC MEASURE

found on the next page, as follows: at the head of each column appears the aesthetic measure; the chordal sequence in question is entered by means of the figures 1, . . . 7, for the respective roots, these being written in Roman or italic form according as a fundamental position or first inversion is indicated. Thus *47* would symbolize IVa → viib.

A set of illustrative chordal sequences in four-part harmony, with corresponding aesthetic measures underneath, is also given in full and follows the table. Here the first line gives the sequences beginning with the tonic chord in fundamental position; the second, those beginning with the first inversion of this chord; the third, those beginning with the supertonic chord in fundamental position; and so on. The first group in each line contains those rated as good by Prout; the second, those rated as possible; the third, those rated as bad.

It will be seen that there is complete agreement as far as could reasonably be expected, with M less than 6 and greater than 9 corresponding generally to the 'bad' and 'good' sequences respectively.

As Prout remarks in this connection (p. 30), "the words 'Good,' 'Possible' and 'Bad' must not be taken as more than mere approximations."

In explanation of the aesthetic measures of the list, it should be remarked that in two cases (*36* and *67*) a chord of measure only 1 was used in order to secure harmonic progression; strictly speaking, only chords of measure at least 2 are to be employed. Furthermore, in the sequences *37*, *57*, the second chord is regarded as the second inversion of V7 rather than as viib, since the dominant is borrowed from the preceding chord; and thus the negative element $BL = 4$ is attributed to these two sequences of the list.

21. General Remarks

The above theory provides a method of theoretic comparison of all possible regular choral sequences, such as would be entirely impossible by means of any set of empirical, negative rules.

The substantial agreement of the theoretical ratings with Prout's table, together with the fact that the theory is based on verifiable elements of order, weighted in a simple way, seems to me to furnish conclusive

DIATONIC HARMONY

evidence in support of the general correctness of the theory. In fact no unfounded theory of this kind could hope to rate correctly 144 types of chordal sequences without important exception, any more than a simple cryptogram could apply to an arbitrary collection of 144 letters.

Undoubtedly the above theory could be considerably improved. However, I doubt whether any modification which takes account of more recent musical forms can be made as yet, since these forms have not become sufficiently well established. Indeed it is entirely conceivable that classical harmony will remain a permanent convenient norm from which interesting stylistic sallies may be made in many directions, but to which there will be a tendency to return.

Aesthetic Measure M and Prout's Ratings of Chordal Sequences

Prout's Rating	Bad						Possible(*)				Good								
$M =$	−1	0	1	2	3	4	5	6	7	8	9	10	11	12	13	14	15	16	17	18
	32	73		54	43	*13*	23	21^b	13*	13*	12^g	12	*12*	*14*	*15*	*14*	*15*	*14*		*17*
	73					23	21	31^b	16*	12^g	13*	16	*16*	*15*	*25*	*14*	*65*	*17*		
						31	34	36*	24*	16*	27^g	24	*21*	*32*	*27*	*15*		*51*		
						34	32	46*	25*	23*	26^g	26	*25*	*41*	*41*	*51*				
						35	37	61*	31*	24*	36^g	26	*21*	*45*	*53*	*54*				
						43	52	64*	34*	26*	32*	23	*24*	*41*		*56*				
							63	61*	35*	36^g	34^g	37	*25*	*45*		*57*				
							65	63^b	42*	35*	31*	42	*43**52*		*51*				
								72*	53*	36^g	46*	47	*42*	*53*		*65*				
								74*	63*	35*	43^g	45	*46*	*54*		*71*				
								75*		45^g	56*	*41**	*52*	*57*						
								75*		46*	62^g	51	*56*	*71*						
										42*	61*	62	*56*							
										47*	64^g	64	*62*							
										54*	72^g	67	*64*							
										52*			*65*							
										53*			*76*							
										63^g										
										61*										
										67*										
										62*										
										76^g										
										74*										

The figures indicate the roots (1 = tonic, etc.): Roman type for fundamental position, italic type for first inversion. Thus *14* stands for Ia → IVb. For $M = 6, 7, 8, 9$, the superscripts g and b indicate 'good' and 'bad' respectively, according to Prout.

(147)

AESTHETIC MEASURE

ILLUSTRATIVE SET OF CHORDAL SEQUENCES
(Major Key)

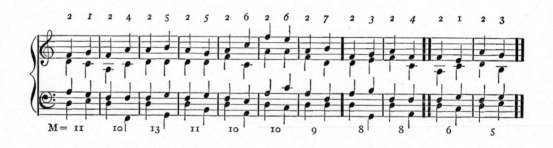

DIATONIC HARMONY

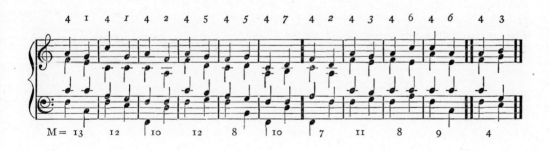

AESTHETIC MEASURE

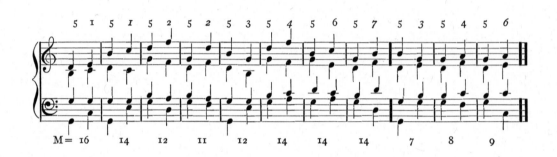

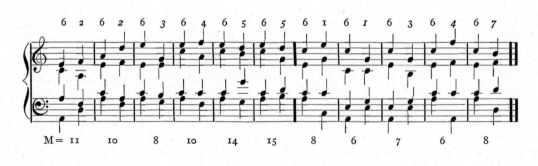

DIATONIC HARMONY

CHAPTER VII

MELODY

1. Introduction

THE two preceding chapters have been devoted to the problem of harmony. The equally important problems of melody and rhythm remain for consideration. We shall concentrate our attention upon melody, although we shall find (section 13) that rhythm is fundamentally a special kind of melody.

No attempt will be made to state any theory of the structure of actual music as dependent upon combined harmony, melody, and rhythm. The simplest and most natural conjecture would be that the aesthetic measure M of any musical composition is given by the ratio O/C, where C is the number of notes which enter as melodic constituents, and O is the sum of the elements of order of harmonic, melodic, and rhythmic types, these being appropriately weighted.

2. The Problem of Melody

In order to separate melody as much as possible from harmony and rhythm, we shall limit attention to simple melodies in one part, made up of 8, 16, 32, 64, or 128 notes, divided into measures of four equal notes. In each measure the first and third notes are to be thought of as theoretically accented. Furthermore, we shall exclude the use of the minor mode and of modulation, by requiring that all of the notes are those of the ordinary diatonic scale.

Despite these stringent limitations, almost all well known melodies may be reduced to this form, or to a like form in three-part instead of four-part time. In the first place most melodies have one predominant melodic part, usually the soprano, which furnishes the essentially equivalent simple melody. In the second place rhythmic and ornamental effects may be eliminated by retaining only the proper principal notes,

MELODY

although frequently this can be accomplished only with definite diminution of musical effectiveness. Moreover, if the melody be longer than 128 notes, there will usually be some shorter characteristic part which underlies the entire melody.

Although we only consider four-part time, it is obvious that the theory here put forth tentatively can be extended easily to other types such as three-part time.

For these reasons the basic problem of melody can be reduced to the following typical simplified form:

Given a simple melody in four-part time of 8, 16, 32, 64, or 128 equal notes of the diatonic scale, to determine a suitable measure of the complexity C and of the elements of order O, such that the ratio $M = O/C$ furnishes a suitable aesthetic measure of the melody.

It is obvious that the number of notes of the melody gives a suitable measure of the complexity, C; of course a note which is held is to be regarded as a single note. Thus the essential difficulty in the problem of melody lies in the effective determination of the elements of order, O.

3. The Rôle of Harmony

A simple (Western) melody is always enjoyed as if it possessed an accompanying harmony, at least to the extent that each of its notes is construed to lie in some definite chord. In case the melody is derived from a musical composition this harmonization coincides with that known through the composition.

On the other hand if the melody is known merely as a tune, a simple automatic mental harmonization will be attached to it. The general law of this automatic harmonization is that it is accomplished by means of chords which are modified as infrequently as possible, with preference for chordal sequences with high aesthetic ratings. Furthermore, in so far as possible, the chordal sequences of the alternate accented notes are to have the same characteristics; the sequence of these accented notes forms what we shall call the 'secondary melody.' In what follows, therefore, we shall assume that the notes of the melody are construed as lying in definite chords.

4. The Question of Phrasing and Comparison

Just as a poem is expressed in lines, so any melody is arranged in musical phrases, often of equal length, which are terminated by a sustained note or an equivalent rest.

Various types of musical phrasing have come to be used just as various types of linear arrangement and rhyming schemes are used in poetry. The simplest type is that in which the melody is divided into four phrases of equal length. This will be the case in most of the illustrations which we shall use, but the theory applies equally well in more complicated cases.

In consequence we shall suppose that any melody under consideration is taken in its usual phrasing. The general character of such phrasing is that the whole melody falls into a certain number of major co-ordinate parts, each of these in turn falls into a number of smaller co-ordinate parts, and so on until the individual phrases are reached. These different types of parts are comparable to the canto, stanza, and line in a poem.

One of the especial advantages of musical phrasing is that it brings out certain comparisons by similarity or contrast between various parts. Presumably the best phrasing and interpretation is that which brings out most clearly the relations of order which exist in the melody.

We shall not attempt to deal with the question as to why one type of arrangement of comparable phrases is more used than another. To do so would be futile, because the accepted forms are largely designated by pure convention; the same kind of arbitrariness prevails in the selection of forms of verse.

One might indeed attempt such an analysis. For example one might ask why the type of phrasing, *AABA*, is preferred to *ABAA*; here *A* represents one phrase and *B* represents a contrasting phrase. Evidently the first arrangement fixes *A* firmly in mind so that the contrast of *B* with *A* is clearly defined. In the second arrangement this contrast is less clearly defined of course, and the *two* final *A*'s seem repetitious since the true function of the concluding *A* is only to effect a return to the starting point. Obviously such an analysis, however valid, falls within the domain of 'qualitative' rather than of 'quantitative' aesthetics.

For these reasons we shall prescribe the musical form used as being

MELODY

established by convention. However, there seem to be certain simple principles which may be stated in this connection:

Any part B made up of one or more phrases may be compared by similarity or contrast with at most one co-ordinate preceding part A. When, however, a part B has been compared with such an earlier part A, no subsequent part is to be compared with this same earlier A, although such a part may be compared with a larger part including A. Within the limits of a single phrase, the second measure may be compared with the first, and the second two measures with the first two, or the second half of the phrase with the first half.

The types of comparison just mentioned are the ones which we shall take account of. They are the principal types which enter into the simple melodies under consideration here. It is hardly necessary to remark that the trained ear appreciates much more elaborate and subtle forms of comparison in complicated musical structures.

5. Limitation of Leaps

Little effort is required when the voice repeats a note or moves by a step upwards or downwards. In a similar way a sequence of notes in a single consonant chord is readily sung provided that the voice does not leap more than an octave. There remain, however, certain types of leaps which are generally avoided in simple melodies. The principal rules of limitation will be formulated as follows:

Dissonant leaps other than by a diminished fifth (from B up to F, or F down to B) or a minor seventh in a dominant 7th chord are forbidden. Two successive leaps in the same direction must either lie in a triadic chord or in a 7th chord; in the latter case the total leap must be a seventh upwards, and the following note must fall a step as in a resolution. Any leap of more than a fifth is to be approached and left in the direction opposite to that of the leap. After two or more steps in one direction a further leap in the same direction must be to an accented note. These rules are essentially empirical ones based on existing practice.* Their reasonableness is obvious.

* Cf. Prout, *Harmony: its Theory and Practice* (31st edition), p. 23.

AESTHETIC MEASURE

6. Preparatory Analysis of a Beethoven Chorale

Let us attempt to specify the principal elements of order as they appear in a classical melody, for instance in the chorale from the fourth movement of Beethoven's Ninth Symphony. In an obvious notation * the melody may be written in the following form

3345	5432	1123	$\dot{3}\widehat{22}$
3345	5432	1123	$\dot{2}1\widehat{\bar{1}}$
2231	23(4)31	23(4)32	$12\widehat{\bar{5}\bar{5}}$
3345	5432	1123	$\dot{2}1\widehat{\bar{1}}$

This melody with a simple harmonization is often used in hymn form.

In the above notation the following simple conventions are employed: An ordinary tie above a repeated note indicates that the note is held; a period above a note indicates that it is lengthened so that the following note is reduced to an eighth note; a dash above or below indicates that the note is an octave above or below the principal octave; a note in parenthesis is one which appears in the complete melody but is treated as an embellishment in the formal analysis.

Evidently there are four phrases of four measures each in the above melody, of which the second is to be compared with the first, the third and the fourth with the first and the second. In fact the form may be indicated as follows:

$$A\,(V),\ A\,(I),\ B\,(V),\ A\,(I),$$

where the first, second, and fourth phrases A are essentially alike save that the first of these has a dominant close (V) while the other two phrases have a tonic close (I). The sharply contrasting phrase B has a dominant close (V). This is a commonplace type of song form.

Let us make an analysis of this melody by successive measures. In doing so various types of elements of order will be brought to light. The reader who wishes to see the complete rule for each item counted is referred to the immediately following section 7.

The first note is the mediant which is in the tonic chord. Now it is a well known fact that most musical compositions commence in the tonic

* 1 = tonic, 2 = supertonic, etc.

MELODY

chord and thus define the tonality. If this expectation of the tonic chord is fulfilled, the fact is appreciated by the ear.* In the case at hand this first element of *tonic start* is present and we assign to it a value 1 as far as the first note is concerned.

The second note is also in the tonic chord so that we add 1 more for tonic start on that account; the third note is not in the tonic chord. This gives a complete count of 2 for this element of order.

Furthermore the second note repeats the first. By taking account of this element of *direct repetition*, for which we assign a value 1 also, we obtain then the count of 3 in all for the first two notes.

The last three notes of the first measure form a melodic sequence. Since the sequence is not established for the ear until the second note is heard, we assign a value 1 to each of the last two notes for this element of *melodic sequence*.

We have now exhausted the obvious relationships of order in the first measure. There remains, however, another element of *harmonic contrast* of a more subtle type. When three notes of a measure lie in a consonant chord, while the fourth one does not lie in that chord, a pleasing harmonic contrast is secured between the notes of the measure. In the case at hand three of the four notes are in the tonic chord. Hence we assign 2 for this harmonic contrast in the first measure, counted 1 for each of the last two notes. Thus the total count for the first measure is 7.

The second measure starts off with the repetition of the preceding dominant note, which adds a count of 1. The four notes of this measure form a falling melodic sequence, for which there is a count of 3 (one for each note after the first). The first three notes of this measure evidently constitute an exact inversion of the last three notes of the first measure. We add a count of 3 for this element of *inversion*, 1 for each note involved. Moreover the middle two notes of the second measure 5432 stand in *melodic contrast* with the corresponding notes of the comparable first measure 3345 (see section 4) from which they differ by step; and so does the last note, since it is not found in the first measure. Hence we assign a count of 1 for each of these notes. Finally, there is a count of 2 because

* In his interesting thesis *The Rôle of Expectation in Music*, New Haven (1921), A. D. Bissell has found that in over 96 per cent of a large number of typical cases the melody does commence in the tonic chord.

AESTHETIC MEASURE

of the *repetition of accented notes* of the first measure. Consequently there is a count of 12 in all for the second measure.

In proceeding to the third measure, we observe that the first note continues the melodic sequence, for which we add a count of 1; the second note repeats the first one and a preceding accented note is repeated, which gives a further count of 2 for repetition; the last three notes form a melodic sequence, for which there is a count of 2; all of the notes but the third are in the tonic chord, for which there is a count of 2 for harmonic contrast; and since the repeated tonic has not appeared before in the first half of the phrase (the two halves of the first phrase are comparable), there is a count of 2 for melodic contrast.

The new element entering in the third measure is that of *transposition*, for obviously the third measure is a precise transposition of the first and is felt as such. We add a count of 4, since four notes are involved in the comparison. Hence the total count for the third measure is 13.

In the fourth measure the transposition continues for two notes more (2); the first note repeats the preceding (1); an accented note of the preceding measure is repeated (1); the first two notes form an inversion of the two preceding notes (2); the last (held) note repeats its predecessor (1); all but one of the notes are in the dominant chord (2); the third note contrasts melodically with the third note of the comparable second measure (1) and the fourth note is the same as the fourth note in the second measure (1). Furthermore there is a count for *cadence* of 1 in the final note, since there is a half cadence in the closing dominant chord of the first phrase. Accordingly the count for the fourth measure is 12.

In proceeding to the fifth, sixth, and seventh measures which repeat the first three, we note first an element of order of 1 for each note because of this *repetition in comparable phrases*. Otherwise, aside from the element of tonic start (2) which is lost here, and the element of repetition of a preceding accented note (1) which is gained, the count is the same as before. Hence the fifth, sixth, and seventh measures give a count of 10, 16, and 17 respectively.

The eighth measure yields a count of 14 as follows, without reference to the first four measures: repetition (3); inversion (2); melodic sequence (2); melodic contrast (3); harmonic contrast (2); moreover there is an

MELODY

authentic cadence (passage from the dominant to the tonic chord) at the end of the second phrase which is counted as 2 more. There is a further count of 4 because of melodic contrast with the fourth measure. Thus the total count for the eighth measure is 18.

In this way we obtain a total count of 44 and 61 in all for the first and second phrases respectively, and so of 105 for the first half of the melody, an average of 3.5 for each of the 30 notes involved.

If we count the third phrase in the same way, we obtain a count of 46 in all. Here the strong melodic contrast with the comparable first phrase is to be observed, for 10 of the 16 notes involved are in such contrast with the corresponding notes of the first phrase. Furthermore the approximate direct repetition of the first measure by the second and of the second measure by the third (all but one note the same in each case) yields a count of $3 + 3 = 6$ for this kind of direct repetition.

The fourth phrase is identically the same as the second with which it is naturally compared, since the first half of the melody is co-ordinate with the second half. Of course this phrase is not to be compared with the third phrase.

The first three measures of this concluding phrase yield the same count as the corresponding measures of the second phrase except that there is a loss of 1 since the first measure contains no accented note of the preceding measure. Thus there is a count of $9 + 16 + 17$ or 42 for these measures. The fourth measure loses a count of 4 because there is no melodic contrast. But there is a gain of 4 because of repetition, and of 3 for *tonic close*. Thus there is a count of $18 + 3$ or 21 for the concluding measure, and so of 63 in all for the last phrase.

This gives a count of 214 in all for the four phrases, an average of about 3.5 elements of order for each of the 60 notes involved.

But, according to the theory here advanced, it remains to add in the count of the *secondary melody* formed by the 32 accented notes. In my opinion there is little doubt that this somewhat concealed melody forms an effective factor in the aesthetic enjoyment. In the particular case under consideration the secondary melody is as follows:

3453	1232	3453	1221
2323	231<u>5</u>	3453	1221

AESTHETIC MEASURE

and possesses obvious melodic quality. The harmonization of this brief melody given below has kindly been effected by my colleague Professor W. H. Piston.

The only new element of order which we find here is that of *harmonic sequence*. Here the harmonic sequences 531, 531, 31$\underline{5}$3, 531 in the tonic chord appear, for which we give a count of one for each note after the first and so of $2 + 2 + 3 + 2 = 9$ in all. The count is 89 for this melody.

Thus the grand total for the Beethoven chorale is 303 or an average of about 5 elements of order for each note. In other words the aesthetic measure M for this Beethoven chorale is about 5, which turns out to be very high in comparison with that of other melodies examined.

7. Definition of O, C, and M for Simple Melody

The following precise but tentative definitions of the elements of order in simple melody are suggested by the analysis of the above melody and of many others.

(1) *Tonic Start and Close*

There is a count of 1 for each note at the beginning as long as these lie in the tonic chord and are in the first measure.

There is a like count of 1 for each note at the end as long as these lie in the tonic chord and are in the last measure; there is a further count of 1 if the last note is the tonic itself.

MELODY

(2) *Cadence*

If there is a passage from dominant to tonic at the close of a phrase (that is, the final change of chord is from dominant to tonic) there is a count of 1 for each note involved and so of 2 in all.

If the final chord is the dominant (half cadence) there is a count of 1 for the final note.

(3) *Repetition of Accented Notes*

According as one or both accented notes of a measure reappear in the following measure there is a count of 1 or 2 as the case may be, provided this is not caused by a mere repetition of the first measure. If both accented notes of the first measure are the same, there is a count of 1 only of course.

(4) *Direct Repetition*

If a single note, or a pair of two notes of which the first is accented, be directly repeated, or a measure, or a larger part not the half of a phrase, be repeated within a phrase, there is a count of 1 for each note of the first repetition.

Moreover the approximate repetition of parts as large as a measure is counted provided there is at most one exceptional note for each measure, the count being 1 for each non-exceptional note; in this case the second repetition is also counted.

(5) *Repetition in Comparable Phrases*

If a part of one phrase is repeated in corresponding position in a later comparable phrase, or if corresponding notes in such a phrase are repeated, there is a count of 1 for each repeated note.

(6) *Transposition*

An exact transposition within a phrase, of at least a measure in length but not all in one direct melodic sequence, counts 1 for each note of the first transposition. If the transposition be repeated a second time within a phrase, there is a count of 1 for each note of the second transposition, provided that the successive transpositions differ by an equal number of degrees.

(7) *Inversion*

In a direct inversion of a rising or falling sequence of at least two notes there is a count of 1 for each repeated note, up to a count of 4.

(8) *Melodic Sequence*

In a rising or falling melodic sequence of at least three notes there is a count of 1 for each note after the first, up to a count of 4.

(9) *Harmonic Sequence*

A harmonic sequence of at least three notes lying in the same consonant chord is counted 1 for each note after the first, up to a count of 4.

(10) *Melodic Contrast*

If a part B is compared (cf. section 4) with an earlier part A, there is a count of 1 for each note of B which either differs by step from the corresponding note of A or which is different from any note found in A. A sustained note is counted as double here.

A phrase B will only be said to contrast with an earlier comparable phrase A in case the count for melodic contrast is at least one half the number of notes in B differing from the corresponding notes of A.*

(11) *Harmonic Contrast*

If all but one of the notes of a measure fall in a consonant major chord, there is a count of 1 each for the last two notes.

(12) *Secondary Melody*

A complete count of the elements of order of the above types is to be made for the secondary melody formed by the alternate accented notes.

The order O is the total count of all the elements of order O of the above types. The complexity C is the total number of notes of the melody. The aesthetic measure M is then the ratio O/C.

---

* Thus the song form $AABA$ is only considered to be followed if B contrasts with the first phrase A in this manner.

MELODY

8. Further Conditions of Satisfactory Form

There are some further conditions which must be fulfilled if satisfactory form is to be achieved. It is almost certain that the conditions of this kind enumerated below are incomplete.

(1) *Ease in Singing*

The rules of section 5 on the limitation of leaps will be adopted.

(2) *Regularity of Pattern*

Some established musical pattern must be followed. This implies a suitable distribution of half cadences and cadences in the successive phrases, as well as a tonic close, so that there is a sense of progression throughout.

(3) *Continuity*

If one or more notes fail to be connected with preceding notes through at least one element of order, there is felt to be discontinuity in the melody, inasmuch as these notes are not suggested naturally by what precedes. Such notes will not be admitted.

(4) *Freedom from Obvious Formal Blemishes*

It must not be possible to increase the total order O by alteration within a short succession of notes (say not more than four), together with corresponding alterations in its repetitions, transpositions, and inversions.

This condition operates to eliminate obvious cacophony because cacophonous repetitions do not lead to counted elements of order according to the rules adopted in section 7.

(5) *Treatment of the Leading Note*

Unless the leading note occurs in a rising or falling dominant sequence of at least three notes, or in a repetition, transposition, or inversion, it must rise to the tonic note directly or *via* the supertonic, or fall to the submediant note.

In fact the leading note so strongly suggests the tonic that unless it is imbedded in a dominant sequence or in a fixed pattern, it must proceed to the tonic note or to the related minor tonic.

(6) *The Secondary Melody*

The secondary melody (aside from the question of conditions (1)–(5) above) must rate as fair; for definiteness we shall require that M shall rate at least as high as 2 for the secondary melody.*

(7) *Rhythmic and Melodic Embellishment*

If a note be directly repeated, it is often desirable to interpret it as a note which is held. Thus an advantageous rhythmic variation may be introduced. Similarly the insertion of eighth notes may soften a considerable leap, or introduce valuable elements of contrast, or increase the rhythmic interest.

Wherever possible we shall introduce such slight embellishments in the final interpretation, just as these were eliminated at the outset for purposes of exact formal analysis. By so doing, obvious awkwardness in the narrow type of simple melodies here treated can be mitigated without any fundamental alteration of the melodic relations involved.

9. THE APPLICATION TO MELODY

The rules contained in sections 7, 8 resulted from the study of a variety of simple melodies. In each case I tried to determine the effective elements of order by the method of introspection. I was kindly aided by Dr. C. B. Morrey in the experimental work.

My principal conclusions may be summarized as follows: (1) for good simple melodies the aesthetic measure M is always high and in general at least 3; (2) it is not possible to find purely mechanical sequences of notes for which M rates as high as 3; (3) sequences cannot be devised with a fairly high rating but devoid of melodic quality. In other words the theory seems to be fairly satisfactory as far as it goes.

The reader who wishes to examine special melodies can readily verify the first conclusion. In order to verify the second conclusion, let him try to construct mechanical sequences of notes with a high aesthetic measure M. In partial justification of the third conclusion I propose to turn next to certain experimentally constructed melodies, devised in the light of the

---

* In obtaining M for the secondary melody it is not considered necessary to treat its secondary melody, that is, the tertiary melody of the original melody, which is formed by the initial notes of the measures.

MELODY

theory and not by use of musical imagination. If these seem tolerable to the ear, the theory is justified to that extent. It is to be stressed that these experimental melodies were the *only ones* which I tried to build, and thus do not represent a selection out of a number of attempts. Furthermore they were constructed in the manner indicated below, and in very brief time indeed. Professor Piston has kindly given these attempts as favorable a setting as possible by supplying an appropriate harmonization.

10. First Experimental Melody

The first experimental melody will be taken to be the secondary melody of the Beethoven chorale (see page 160).

In order to rate it we must add to the count 89 already obtained a further count for the elements of order found in its secondary melody:

$$3513 \qquad 3512 \qquad 2221 \qquad 3512$$

The total of these is readily verified to be 39, so that the total count is 128, and the aesthetic measure M of the first experimental melody is 4 as against 5 for the Beethoven chorale from which it was derived.

11. Second Experimental Melody

The second experimental melody was obtained as follows: The Westminster chimes melody is essentially the following little melody of 16 notes in four equal phrases:

$$31 2\underline{5} \qquad \underline{5} 231 \qquad 321\underline{5} \qquad \underline{5} 231$$

in which the last note of each measure is accented. The count for the elements of order is 49. In the secondary melody

$$32\underline{5}3 \qquad 31\underline{5}3$$

the count is 18.* Thus O is 67 and M is 4.2.

The first step was to construct an experimental melody of 32 notes of which the secondary melody is precisely the above Westminster chimes melody. This was done in the following manner:

$$\begin{array}{cccc} 3211 & 2\underline{755} & \underline{5}123 & 3212 \\ 3322 & 11\underline{55} & \underline{5}123 & 3212 \end{array}$$

* This tiny melody of eight notes is taken as made up of two measures, each of which forms a phrase.

AESTHETIC MEASURE

Here the precise selection of the non-accented notes was largely determined by the following considerations. The Westminster chimes melody suggests a four-part song form *ABCB* in which *B* and *C* contrast with *A*. In the first measure it was obviously easy to fill up 3∗ 1∗ to be 3211, and thus introduce the elements of melodic sequence, repetition, and harmonic contrast.

In the second measure 2∗ 5∗ the leap from 2 to 5 was most easily made by introducing an intermediate 7; by completing this measure to be 2755 the four notes were placed in a harmonic sequence within the dominant chord, and the second measure stood in complete melodic contrast to the first, since its first note differed by step from that in the first measure, while the other notes appeared in the second measure for the first time.

Similarly if we try to fill out the third measure 5∗ 2∗ so as to contrast with 3211 (the third note being in contrast) we are naturally led to 5123, especially since 2 is accented in the preceding measure and harmonic contrast is thereby secured.

In the fourth measure 3∗ 1∗, there is no possibility of securing effective melodic contrast with the comparable second measure 2755. The obvious completion to 3212 has evident advantages when taken in conjunction with the first and third measures; in particular the final note gives a half cadence.

We shall not remark upon the second half of the completion, except to observe that the third phrase is chosen in as strong melodic contrast to the first as possible, while the fourth phrase is taken as identical with the second in conformity with the song form selected.

In an entirely similar manner the above melody of 32 notes was treated as a secondary melody and expanded once more to the following melody in four phrases:

3323	1211	2176	5655
5112	2334	3323	1122
3236	2322	1714	5655
5112	2334	3323	1211

This is given opposite in the key of G with the harmonization by Professor Piston. The sustained notes have been introduced so as to relieve the rhythmic flatness.

MELODY

The total count for the elements of order in this melody is 121, while the count for the secondary melody is 88. Consequently the aesthetic measure is 209/58 = 3.6. Hence this is a fairly satisfactory melody, according to the theory given above. The reader will have to judge for himself whether it actually is so or not for him.

12. Third Experimental Melody

The third melody was devised in the course of an hour by taking as the point of departure the measure 3236 which is the ninth measure of the preceding experimental melody. It was constructed step by step, with the above definition of M in mind, and, in particular, with due regard to the secondary melody. As I first wrote it, the twelfth measure was $\overline{1233}$; Professor Piston suggested the modification to $\overline{1766}$. This is obviously a definite melodic improvement which not only avoids the leap of an octave at the end of the twelfth measure, but gives the more acceptable phrase form

$$A\;(V),\;A\;(I),\;B\;(vi),\;A\;(I),$$

whereas the first form

$$A\;(V),\;A\;(I),\;B\;(I),\;A\;(I)$$

AESTHETIC MEASURE

is not usable because all of the three last phrases end in the tonic chord. It may be remarked in passing, however, that although the original song form is not an acceptable one, the sequence $\overline{1233}$ yields a somewhat higher count.

The precise form of the experimental melody arrived at in this manner was the following:

3236	55$\overline{1}$5	4321	71$\overline{22}$
3236	55$\overline{1}$5	4321	12$\overline{11}$
4455	6677	$\overline{1232}$	$\overline{1766}$
3236	55$\overline{1}$5	4321	12$\overline{11}$

The harmonization supplied by Professor Piston in the key of E^b is given herewith.

The total count for the elements of order in the above melody and its secondary melody is $165 + 74 = 239$ so that the aesthetic measure M is about 4. This experimental melody is clearly more successful than the preceding one. In estimating its quality the reader should bear in mind that only melodies without rhythmic effects are being considered.

MELODY

13. THE PROBLEM OF RHYTHM

Concerning rhythm we propose only to make the following fundamental remark:

The important problem of rhythm is akin in its essential nature to the problem of melody.

This fact may be seen as follows. Let us imagine that only a full beat A and two half beats B are to be used as units. Any rhythmic pattern based on these then defines a sequence at equal intervals of time such as

$$BABA \quad BBBA \quad BABA \quad BBBA \quad \text{etc.}$$

This may be regarded as a kind of melody in the two notes A and B. We shall not develop this analogy further.

The problem of rhythm should be carefully studied from the point of view of the theory of aesthetic measure; it is essentially simpler than the problem of melody which we have considered.

CHAPTER VIII

THE MUSICAL QUALITY IN POETRY

1. THE TRIPARTITE NATURE OF VERSE

IN the analysis of poetry we shall make a division of the aesthetic factors into those pertaining to significance, to musical quality, and to metre. The significance of a poem is to be found in its 'poetic ideas,' which must be expressed in accordance with Hemsterhuis's general dictum of the "greatest number of ideas in the shortest space of time." In other words the poet must begin with a poetic vision, and then by the use of poetic license, effective figures of speech, and onomatopoetic devices obtain an adequate embodiment of this vision in terse form; that is, in a form much shorter than would be necessary for an equally adequate expression in the language of prose.

One can find frequent appreciations of this first factor in poetry. The poet Shelley when he defined poetry as "the expression of the imagination" was alluding to the primacy of the poetic idea. The well known important rôle of poetic freedom as an essential element in expression is indicated by Pope as follows:

> Thus Pegasus, a nearer way to take,
> May boldly deviate from the common track.

The terseness of poetry is stressed by Voltaire: "Poetry says more and in fewer words than prose."

This factor of significance is evidently essentially connotative in its nature and beyond any possibility of formal analysis. It is perhaps the most important single element in poetry, and yet poetry without musical quality and metre is not properly poetry at all.

The other two factors of musical quality and metre are much more formal in their nature. It is because of these factors, which correspond to those of harmony and rhythm in music, that Fuller wrote "Poetry is music in words, and music is poetry in sound."

THE MUSICAL QUALITY IN POETRY

We shall say nothing of metre in what follows except to mention its similarity to rhythm in music. This similarity has been considered by Lanier in his *Science of English Verse,* who goes so far as to use musical notation in order to specify metre. It is to be expected that metre in poetry will be susceptible of mathematical analysis in regard to aesthetic effect, just as is rhythm in music. As far as metre is concerned, the poet must first select an appropriate and sufficiently pliable metric form, and then in the inevitable deviations from its rigid execution be guided by a delicate sensibility of the effect upon the ear.

The remaining factor of musical quality, to which we devote our attention, is in certain respects the most characteristic one. Thus Butler wrote:

> For rhyme the rudder is of verses
> With which, like ships, they steer their courses.

And Shelley lays a similar emphasis: "The language of the poets has ever affected a sort of uniform and harmonious recurrence of sound, without which it were not poetry and which is scarcely less indispensable to the communication of its influence than the words themselves without reference to that peculiar order." It is here that the ingenuity of the poet is much more exercised than in following some more or less arbitrary metric form.

2. THE MUSICAL QUALITY IN POETRY

It must be borne in mind that the notion of musical quality, as separable from significance and metre in a poem, is only approximately tenable. In such a line as the following of Tennyson's:

> The league-long roller thundering on the reef,

there is expressed an imaginative theme for the eye and ear in onomatopoetic language of the utmost terseness, having appropriate metric structure and unusual musical quality. This line is, however, to be looked upon as produced by the intimate union of these factors, and so as being much more than a mere aggregation of them all. Nevertheless it is certain that the musical factor is to a large extent appreciated by itself, so that two poems can be intuitively compared in regard to their musical quality, almost regardless of their significance or metric form.

3. Rhyme

The first and most obvious of the simple musical factors in poetry is rhyme. Here, in Western poetry, one group of sounds is compared with another in the following manner. The initial (elementary or composite) consonantal sound of the first group corresponds to a distinct initial consonantal sound of the second group; the remaining sounds of the two groups are identical. It is required furthermore that the two groups contain one accented vowel sound having the same relative position in the two groups, and that these groups terminate with a word. The following are simple instances of such rhyming groups: Kh*an*, r*an*; decr*ee*, s*ea*; f*ire*, des*ire*; n*umbers*, sl*umbers*; rem*ember*, Dec*ember*. There are also certain slight licenses that are admitted in rhyme, to which we can only allude in passing.

The element of rhyme in poetry is analogous to the element of melodic contrast in melody, and plays a considerable part.

4. Assonance

Under certain circumstances the repetition of a vowel sound gives rise to a pleasurable feeling of assonance; it is only in this narrow sense of vowel repetition that we shall employ the word 'assonance.' The effect of assonance is increased by further repetitions, at least up to a certain point, after which the excess of assonance becomes unpleasantly monotonous and disagreeable.

The factor of assonance in poetry is analogous to that of repetition in melody. Its use is illustrated, for instance, by the following opening lines from Poe's poem, 'The Bells':

> Hear the sledges with the bells —
> Silver bells!
> What a world of merriment their melody foretells!

These two lines contain the vowel sound of *e*, as in 'bells,' eight times in twenty-three syllables, and yet this repetition is not felt as excessive. Throughout Poe's poem the same sound is repeated extremely often. This repetition would be felt to be monotonous if the recurrence did not contain an onomatopoetic suggestion appropriate to its subject, 'The Bells.'

THE MUSICAL QUALITY IN POETRY

5. Alliteration

By 'alliteration' we shall mean only such repetition of consonantal sounds as is felt by the ear. For instance, in the first two lines from 'The Bells' there is alliterative occurrence of the sound $s = z$.

A more nearly excessive use of alliteration and assonance occurs in the following interesting illustrative stanza given by Poe in his essay, 'The Rationale of Verse':

> Virginal Lilian, rigidly, humblily dutiful;
> Saintlily, lowlily,
> Thrillingly, holily,
> Beautiful!

Here the consonant sound of l appears no less than sixteen times in the thirty syllables while the short vowel sound i occurs thirteen times. Observe also the five times occurring repeated syllable, *lili*. Of course a fundamental flaw in this stanza is the appearance in it of two non-existent adverbs, 'saintlily' and 'lowlily.'

Alliteration, like assonance, is analogous to the element of repetition in melody.

6. The Musical Vowel Sounds

In general the vowel sounds are smoother than the consonantal sounds; and among the vowel sounds there are certain ones which are especially musical in quality, notably the *a* as in *a*rt, the *u* as in t*u*nef*u*l, bea*u*ty, and the *o* as in *o*de. When these appear sufficiently frequently, they impart their soft musical character to verse.

The mathematician Sylvester in his *Laws of Verse*, to which we shall refer subsequently, has a footnote of interest in this regard: "I can not resist the temptation of quoting here from a daily morning paper the following unconsciously chromatic passage . . . : 'The last portion of the shadow of the earth has been passed through by the moon which then again sailed in its full orb of glory through the dark blue depth.'" Besides possessing pleasant alliterative and assonantal elements, this sentence contains ten musical vowels as follows: l*a*st, sh*a*d*o*w, p*a*ssed, thr*ou*gh, m*oo*n, f*u*ll, gl*o*ry, thr*ou*gh, d*a*rk, bl*u*e. The aesthetic element introduced in poetry by these musical vowels may be compared with that due to the primary chords in melody.

There are doubtless certain special instances where the play of vowels and consonants produces a kind of musical tune in a poem, analogous to melody. Lanier (*loc. cit.*) says in this connection: "Tune is . . . quite as essential a constituent of verse as of music; and the disposition to believe otherwise is due only to the complete unconsciousness with which we come to use these tunes . . . in all our daily intercourse by words."

However, it would be very difficult to disengage this element in precise form, and it is not certain how far its rôle is really independent of that of the musical sounds, alliteration, and assonance, already taken account of. For these reasons we do not attempt to deal separately with this somewhat obscure element of 'tune' in a poem.

7. 'Anastomosis'

Another aesthetic factor which is agreeable is that which results from the fact that a poem is easily spoken. In general this is brought about when the consonantal sounds are simple rather than composite, and not much more numerous than the more easily pronounced vowel sounds. Sylvester used the term 'anastomosis' to express this desirable quality (*loc. cit.*): "Anastomosis regards the junction of words, the laying of them duly alongside of one another (like drainage pipes set end to end, or the capillary terminations of the veins and arteries) so as to provide for the easy transmission and flow of the breath . . . from one into the other."

From our point of view this factor is not a unitary one, but needs to be split up into two others. On the one hand the composite consonantal sounds increase the complexity, C, of the poem, which, in accordance with our theory, is correlated with the effort required in speaking it. Such complexity is heightened also when a word ends with a consonantal sound and the following word begins with a sufficiently different consonantal sound. Hence such sequences are to be avoided as far as possible.

On the other hand, quite regardless of this factor of complexity, an excess of consonantal sounds is felt to be decidedly harsh and disagreeable. For example in the catch:

> Midst thickest mists and stiffest frosts,
> With strongest fists and stoutest boasts,
> He thrusts his fists against the posts,
> And still insists he sees the ghosts —

there are approximately twenty-four consonantal sounds as against only eight vowel sounds in the first line, and a similar relationship holds in the other three lines. Here we have serious consonantal excess, a negative aesthetic factor.

Thus, from our point of view, 'anastomosis' is secured by avoidance of consonantal excess and by the diminution of 'complexity' as far as possible.

8. Poe's Concept of Verse

We have now alluded to the principal elements of order involved in the musical quality of verse. While there exists, so far as I have discovered, no formulation but my own of its 'aesthetic measure,' nevertheless both Poe and Sylvester come within striking distance of an analysis of musical quality and metre in verse. The following quotations from Poe (*loc. cit.*) indicate sufficiently his point of view:

> *Verse* originates in the human enjoyment of equality, fitness. To this enjoyment, also, all the moods of verse — rhythm, metre, stanza, rhyme, alliteration, the *refrain*, and other analogous effects — are to be referred.
>
> The perception of pleasure in the equality of *sounds* is the principle of *Music*.
>
> Anyone fond of mental experiment may satisfy himself, by trial, that, in listening to the lines, he does actually, (although with a seeming unconsciousness, on account of the rapid evolution of sensation,) recognize and instantly appreciate (more or less intensely as his ear is cultivated,) each and all of the equalizations detailed. The pleasure received, or receivable, has very much such progressive increase, and in very nearly such mathematical relations, as those which I have suggested in the case of the crystal.

Evidently in Poe's statements there is an assertion of the quantitative, additive character of the aesthetic pleasure produced, which corresponds closely with our concept of the order, O.

The notion of the complexity, C, does not enter explicitly in his analysis, but doubtless in the background of his ideas is the implicit requirement of a large *density* of such relations of 'equality' as Poe termed them. By equality Poe had in mind not only identity in sound but also metric equality, so that, in effect, he proposed to deal with both mathematical factors of metre and musical quality. He refers nowhere to the factor of significance, although of course it is also implicit in the background of his ideas. It is interesting that, precisely in the respect of significance, Poe fails to be of the first rank among poets.

9. Sylvester's Concept of Verse

There is no doubt that Sylvester's concept of verse was much influenced by that of Poe. However, Sylvester went further than Poe in approaching the point of view demanded by our mathematical theory, as the following quotations show (*loc. cit.*):

> In poetry we have sound, thought, and words . . . ; accordingly the subject falls naturally into three great divisions, the cogitative, the expressional, and the technical; to which we may give the respective names of Pneumatic, Linguistic, and Rhythmic. It is only with Rhythm that I profess to deal. This again branches off into three principal branches — Metric, Chromatic and Synectic.
>
> I touch very briefly on this branch [of Metric] accepting, in regard to it, the doctrine of Edgar Poe. . . .
>
> Metric is concerned with the discontinuous, Synectic with the continuous, aspect of the Art. Between the two lies Chromatic, which comprises the study of the qualities, affinities and colorific properties of sound. Into this part of the subject, except so far as occasionally glancing at its existence and referring to its effects, I do not profess to enter. My chief business is with Synectic.
>
> This, also, on a slight examination, will be found to run into three channels — *Anastomosis, Symptosis*, and between them the main flood of *Phonetic Syzygy*.

Evidently Sylvester's 'Pneumatic' and 'Linguistic' fall under what we have termed the significance (with appropriate brief expression), while his 'Rhythmic' embraces both musical quality and metre. Furthermore he is concerned with musical quality rather than metre.

His last statement is the important one. As we have observed, his principle of anastomosis corresponds roughly to the requirement of as little complexity, C, as possible. On the other hand his principle of symptosis "deals with rhymes, assonances (including alliterations so-called), and clashes (this last comprising as well agreeable reiterations, or congruences, as unpleasant ones, i.e. jangles or jars)." It involves then the same elements of order as are classified above. Consequently symptosis is more or less the counterpart of our order, O.

THE MUSICAL QUALITY IN POETRY

Thus his 'Phonetic Syzygy' — the effective combination of 'Anastomosis and 'Symptosis' — corresponds in a qualitative sense to our aesthetic measure of musical quality. It does not appear that Sylvester believed the aesthetic effect of poetry to be quantitatively measurable as Poe had conjectured it to be.

10. On Phonetic Analysis

As a first step towards the formulation of an aesthetic measure of musical quality in poetry, it is necessary to introduce certain agreements, as simple as possible, concerning phonetic analysis. We shall recognize only the following vowel sounds as distinct: *a*id, *a*dd, *a*rt; *e*ve, *e*ll; *i*sle, *i*ll; *o*de, *o*r; *u*se, *lu*ne, f*u*ll, l*u*ll, *u*rn; *oi*l; *ou*t. All vowel sounds distinct from these are to be assigned to the nearest one of these sounds.

The sounds of *a* as in *a*rt, *u* as in t*u*nef*u*l, bea*u*ty, *o* as in *o*de, will be called 'musical' because of their agreeable quality, similar to that of a pure musical note.

The elementary consonantal sounds are taken to be the following:

$$b, d, f, g, h, j, k, l, m, n(g), p, r, s, t(h), w(h), y, z.$$

The ordinary *g* sound is not regarded as present in *ng* although *n* is present. Likewise the aspirate *h* is not regarded as present in *th* and *wh* although *t* and *w* are present. Neither *c* nor *h* are regarded as present in *ch*, nor *h* in *sh*; all other composite consonant sounds will be analyzed in the manner indicated by the usual spelling. The phonetic justification of these simple conventions is obvious. Vowel and consonantal sounds not used in the English language need not be considered of course, since we are dealing with English poetry only.

11. The Complexity C

The complexity C of any part of a poem — as, for instance, of a single line — is simply the total number of elementary sounds therein, increased by the number of word-junctures involving two adjacent consonantal sounds of the same line, which do not admit of liaison. The following pairs of distinct consonantal sounds will be taken to admit of liaison: $b, p; d, t; f, v; g, k; m, n; s, z$. Likewise if the second sound is an aspirate h, liaison will be admitted.

The justification for the definition is self-evident: We may with fair approximation to the facts consider each vowel sound and elementary consonantal sound as being equally difficult to pronounce. Furthermore, adjacent but distinct consonantal sounds belonging to different words are nearly as difficult to pronounce as if there were a vowel sound between them. The definition takes account of all these facts in a simple way.

12. The Element $2r$ of Rhyme

The number of sound groups in the line or part of the poem under consideration, which rhyme with at least (and in general only) one sound group earlier in the same line or in an earlier line will be designated by r. The corresponding element of rhyme will be taken to be $2r$. It is seen then that the index 2 is assigned to the tone of feeling due to a single rhyme. Later we shall assign an index of 1 to the single alliteration or assonance. The higher rating of the rhyme is justified by the greater intensity of the tone of feeling which each rhyme induces.

The reason for counting each rhyming group only once, even though it rhymes with more than one earlier group, is that each rhyming group is in general set against one particular contrasting earlier group. By general consent certain slight liberties are allowed in rhyming. For instance the u sounds as in *u*se and l*u*ne, are allowed to replace one another. We shall not undertake to enumerate these exceptions.

The evaluation of the element r is immediate in all cases. For example in the case of Poe's stanza there are the following rhyming groups:

low*lily*, ho*lily*; d*utiful*, b*eautiful*.

Thus r is 2 and the element $2r$ of rhyme is 4 in this case.

13. The Element aa of Alliteration and Assonance

In dealing with the element of alliteration and assonance, it is necessary first to specify limits within which the repetition of a consonant or vowel sound is pleasantly felt, and then to decide in how far repetitions may accumulate before there is alliterative or assonantal excess. The short nondescript vowel sound as in th*e*, attent*i*on, which may be considered to be a short form of u as in l*u*ll, will not be regarded as assonantal under any circumstances.

THE MUSICAL QUALITY IN POETRY

In order to give an empirical definition of the element aa of alliteration and assonance, we shall introduce certain technical terms. A 'leading sound' of a line will be defined as one which either is part of an accented syllable, or occurs as the initial sound of a line, or is part of a rhyming group. Such leading sounds are those which most impress the ear.

A sound, a, in a certain position will be said to be 'directly connected' with the same sound or sounds in other positions as follows: (1) the same sound is found again in the same word as a or in an adjoining word; (2) the same sound occurs as a leading sound earlier in the line than a, or in the last half of the preceding line provided a occurs in the first half of its line; (3) a follows a pair of the same sounds earlier in the line which are in the same word or in adjoining words; (4) the same sound occurs as a rhymed leading sound or as a leading sound in the same relative position as a, either in the preceding line or in an earlier line rhyming (or identical) with the line containing a; (5) the same sound occurs in an identical syllable earlier in the line than a or in the last half of the preceding line provided a occurs in the first half of its line.

If there is an odd number of feet in a line, the middle foot will be considered to be part of both the first and second halves of the line in applying these definitions; moreover, if a line contains only two feet, we shall include both feet in either half of the line, by special convention.

When a sound is repeated under these conditions, the repetition produces an effect of alliteration or assonance which is felt agreeably.

With these definitions in mind we shall define the element aa of alliteration and assonance in the line or group of lines under consideration as the number of sounds directly connected with others in the same or preceding lines, but not directly connected with more than two leading sounds or more than four sounds in all.

The reason for the restriction imposed lies in the obvious fact that beyond a certain point alliterative or assonantal play upon a particular sound is felt to be monotonous and even disagreeable. This is in accordance with the usual effect of undue repetition. The precise limits assigned are of course a matter of somewhat arbitrary choice.

In illustration of the above rule let us consider Tennyson's line already quoted in section 2. The leading sounds are evidently the sounds itali-

AESTHETIC MEASURE

cized below, where the figures 1, 2, 3, 4, 5 are set respectively above the sounds *t, l, e, n, r*; these are the only ones to enter in *aa*:

$$\begin{matrix} 1 & 23 & 2\ 4\ 5\ 2\ 5\ 1 & 4 & 54 & 4\ 1 & 53 \\ \text{The \textit{league}-long \textit{roller} \textit{thun}dering \textit{on} the \textit{reef}.} \end{matrix}$$

Here the second and third sounds *t* count since they are directly connected with its initial position; all three sounds *l* are directly connected with one another and are counted; the second sound *e* (long) counts since it is directly connected with the first; the last three sounds *n* count, being directly connected with one another, but the first does not count; the four sounds *r* count since the first two are directly connected, and the last two are directly connected with the first two. Thus *aa* is 13 for this line.

A practical method of evaluation of the element *aa* is to start with the first sound to occur, and put a dot over all its positions which are counted in *aa*, then to do the same with the second sound, and thus continue to the end. The total number of dots gives the element *aa* required, at least unless it happens that the limit of desirable alliteration and assonance is exceeded.

An example of a case in which the limit of desirable alliteration and assonance is exceeded is furnished by Poe's stanza. In such rare complicated cases it is convenient to put small circles above the sounds such as *l* and *i* which are not to be counted in *aa*, but otherwise to proceed as before. Thus we obtain the following:

Virginal Lilian rigidly, humblily dutiful;

Saintlily, lowlily,

Thrillingly, holily,

Beautiful!

It is to be noted that, in accordance with our rule, the rhyming sounds in 'holily' and 'beautiful,' other than the initial *h* and *b*, are to be counted in *aa*. Furthermore the final *y* occurring five times is interpreted as a long *e* and not as a short *i* sound. We find then that *aa* is 39 in this case.

THE MUSICAL QUALITY IN POETRY

14. THE ELEMENT 2*ae* OF ALLITERATIVE AND ASSONANTAL EXCESS

According to the definition of the element *aa* of alliteration and assonance, repetition of a sound beyond a certain point does not have further favorable effect. We have still, however, to make allowance for the fact that alliteration and assonance may become positively unpleasant under certain circumstances.

These circumstances are of the following three types: (1) the sound in question is a leading sound directly connected with too many earlier leading sounds; (2) the sound forms part of successive identical syllables not in the same word; (3) the sound in question is one of a uniform series of regularly recurring repetitions.

The effect produced by the immediate repetition of syllables as in (2) is cacophonous. It was to avoid this fault that Collins changed the second line* of his 'Ode to Evening' beginning

> If aught of oaten stop, or pastoral song
> May h*ope, O pens*ive* Eve*, to sooth thine ear—

to

> May hope, chaste Eve, to sooth thy modest ear.

The element of cacophony may be compared to that of false cadence in harmony: more precisely, the immediate repetition of syllables ordinarily occurs either within a word, or in the agreeable repetition of a word; consequently when such repetition is not of one of these types, there is an unpleasant feeling of artificiality.

Likewise the effect produced by the same sound repeated several times at uniform intervals is artificial and unpleasant, whether these sounds occur in successive words, feet, lines, or stanzas.

Thus our definition of *ae* will be the number of leading sounds directly connected with more than two earlier leading sounds, and of sounds belonging to syllables immediately following the same syllables, but not in the same words, and of sounds belonging to a uniform series of the same sounds which contains at least three earlier sounds. The negative element

---

* See J. L. Lowes, *Convention and Revolt in Poetry* (1919).

of alliterative and assonantal excess will then be defined to be $2ae$. In other words an index -2 will be assigned to the corresponding effect.

It need hardly be said that there will rarely be found excessive alliteration or assonance of this sort in satisfactory poetry, and that when it is present, the ear will note the effect at once. Thus ae is 5 in the case of Poe's stanza because the last four ineffective elements in aa count in ae and the repeated sound *ly* occurs at the end of four successive words.

On the other hand, in the catch of section 7 the first line contains the sound *s* five times as a leading sound and the sound *t* four times as a leading sound, all after the first of these directly connected with the first. Hence, according to definition, two of the sounds *s* and one of the sounds *t* count in ae, so that the element $2ae$ is 6 in this line alone.

15. The Element $2m$ of Musical Vowels

We shall define the element of musical vowels as $2m$ where m is the number of musical vowels (*a* as in *a*rt, *u* as in t*u*neful, bea*u*ty, and *o* as in *o*de) increased by the number of vowels *o* as in *o*r which are directly connected with an earlier long musical vowel *o* or *a*. The limitation will be imposed, however, that such musical vowels directly connected with more than two other earlier vowels will not be counted in m. The limitation is introduced because repetition of musical vowels is of no interest beyond a certain point.

The reason for counting the short *o* after the long *o* is merely that when one tries to pronounce the *o* in such a word as 'or' so that it is long, it tends to take the short form. Thus the long and short forms are closely connected, and if the long musical form of *o* precedes the short form and is not too far from it, the short *o* takes on the same musical quality. Furthermore the closeness of the two sounds justifies the similar rule for the sound *a*.

Such euphonious lines as

> Little boy blue, come blow your horn,

and

> Come into the garden, Maud,

with $2m = 8$ and $2m = 6$ respectively, show clearly the effectiveness of the musical vowels when used in this manner.

THE MUSICAL QUALITY IN POETRY

16. The Element $2ce$ of Consonantal Excess

In case there are in all more than two elementary consonantal sounds for each vowel sound in any line, an appreciably harsh effect is produced. For this reason we shall define the consonantal excess as $2ce$ where ce is the excess of the consonant sounds over twice the number of vowel sounds in each line. In illustration we consider the first line of the catch of section 7. Here there are eight vowel sounds and twenty-four consonantal sounds. Hence the excess $2ce$ in this line alone is 16. The element $2ce$ enters as a negative element of course.

17. The Aesthetic Formula

The complete aesthetic formula for musical quality in poetry is taken as follows:

$$M = \frac{O}{C} = \frac{aa + 2r + 2m - 2ae - 2ce}{C}$$

Here all the elements which enter have been explicitly defined in the preceding sections.

In order to evaluate M systematically, the following method is convenient:

(1) Determine C by direct phonetic analysis of each line. The successive numerals 1, 2, 3, . . . may be placed under the successive sounds of the line and under the junctures not capable of liaison.

(2) Consider the successive sounds in their order of appearance and put a dot over all those which are alliterative or assonantal, and then place a circle around those dots (if any) for which the sound does not count in aa.

(3) Place two additional dots over the accented vowel sounds rhyming with the same sound in an earlier position.

(4) Place two additional dots over each musical sound and a circle around those dots (if any) for which the sound does not count in m.

(5) The total number of dots not enclosed by circles then gives the sum $aa + 2r + 2m$.

(6) Determine the sum $2ae + 2ce$, which is in general 0 in any satisfactory poem.

(183)

AESTHETIC MEASURE

(7) Subtract this sum from that specified in (5), obtaining O.

(8) The aesthetic measure M is then O/C.

In the tabulation of these items it is convenient to put C for each line (step 1) to the right of and below the line, and O (steps (2)–(7)) to the right of and opposite each line. The ratio M giving the aesthetic measure of the line is thus displayed to the right. For a group of lines, the constituent numbers C for each line should be added to give the total C, and the constituent numbers O should likewise be added to give the total O. The aesthetic measure of the lines considered is the ratio of the total O to the total C.

18. Analysis of Five Verses of 'Kubla Khan'

Let us consider a particular application of this rule to the first stanza of Coleridge's 'Kubla Khan,' which affords a very remarkable example of musical quality. The analysis is indicated by the following tabulation:

$$M = \frac{87}{105} = .83$$

```
  .        .            .       .
 . .....   .   . .      ...    . .
In Xanadu did   Kubla Khan              22
1 23456789 10 11 12 13 14 15 16 17 18 19  20 21    21

 .  ...      .      ..        ...
A stately pleasure-dome   decree:        15
1  2345 67 8 9 10 11 12 13  14 15 16 17 18 19 20 21 22   22
                                        .
 .  .          ..     .     . ..
Where Alf, the sacred  river, ran        12
1   2 3   4 5 6 7  8 9 10 11 12 13 14  15 16 17 18 19 20 21 22   22
                                  .
 .  .    .. ...       ... .
Through caverns   measureless
1   2  3    4 5 6 7 8 9 10 11 12 13  14 15 16  17 18 19 20
  .   .
 .. ...
to  man                                  24
21 22 23 24 25                           25
  .           .
 ..   .  .... ..
Down to a sunless sea.                   14
1 2   3 4 5 6 7 8 9 10 11 12 13  14 15   15
```

(184)

THE MUSICAL QUALITY IN POETRY

These computations indicate that the number of elements of order ($O = 87$) is not much less than the number of sounds ($C = 105$), their ratio giving an aesthetic measure of .83 for the stanza. From this point of view the first, fourth, and fifth lines are the most musical.

Lamb made a very trenchant estimate of this remarkable poem in referring to it as "a vision, 'Kubla Khan,' which said vision he [Coleridge] repeats so enchantingly that it irradiates and brings heaven and elysian bowers into my parlour while he sings or says it; but there is an observation, 'Never tell thy dreams,' and I am almost afraid that 'Kubla Khan' is an owl that won't bear day-light. I fear lest it should be discovered by the lantern of typography and clear reducing to letters no better than nonsense or no sense." In other words, the significance of Coleridge's poem is elusive and slight, although the element of musical sound is almost magical. The same criticism would apply to much of Poe's poetical work of course.

19. An Experimental Poem

In his essay on 'The Philosophy of Composition' of 1846, Poe analyzed step by step the construction of his poem 'The Raven' and claimed that his theory had been used as a conscious and effective tool in its composition. What is remarkable here is not that Poe had a theory. Almost every creative artist has a theory or point of view which, for him, sums up the inner secret of his success. Rather it is the fact that Poe expressed his theory in mechanical terms.

As of some interest here I shall give an account of a somewhat similar experiment made by myself on the basis of the theory described above. This experiment was undertaken in order to clarify my own ideas about the nature of poetic composition and to subject them to a test. The reader will have to judge for himself as to the success of the experiment.

According to the theory it was first of all necessary to start from an idea having some poetic quality. Here I chose an idea concerning the general nature of knowledge which I had expressed in prose as follows:*

We may compare if we will, our bits of knowledge to luminous threads which we wind into a compact, luminescent ball. By skilful arrangement of the threads

* Century magazine, June, 1929.

there begins to appear in the center of this ball a bright vision of concepts and laws. If now we add further irrelevant threads, the vision is obscured; and if we unwind the threads in an effort to approach the vision more intimately it becomes more and more faint, and finally disappears.

My first attempts to incorporate this idea in poetical form were very unsuccessful. The chief reason for the initial lack of success seems to me now to lie in the fact that the expression of the idea was not sufficiently *terse*. The requirement of terseness is of course fundamental.

Then one day came without apparent effort the following:

Vision

Wind and wind the wisps of fire,
Bits of knowledge, heart's desire;
Soon within the central ball
Fiery vision will enthrall.

Wind too long or strip the sphere,
See the vision disappear!

The aesthetic measure of this short poem according to the criterion above is .62. Comparison with the ratings of an arbitrarily selected list of poetic lines (section 20) indicates that the poem may be considered as of good musical quality according to our theory. In this case the poetical form of expression, although more terse, falls short of the prose form in exactitude, but has perhaps the advantage of inducing more emotional interest. In the writing of these six lines there was certainly no conscious use of the formula. Nevertheless I believe I could not have done nearly so well without conscious reflection concerning the aesthetic factors in musical quality taken account of by the formula.

20. Further Examples

For the purpose of testing the theory a number of characteristic opening lines of varying musical quality were selected, and then arranged by others in the order of their aesthetic preference, as far as musical quality was concerned. The arrangement thus obtained was found to be substantially in accord with that indicated by the theory as tabulated below.

$$M = .83$$

In Xanadu did Kubla Khan
A stately pleasure-dome decree:

THE MUSICAL QUALITY IN POETRY

Where Alf, the sacred river, ran
Through caverns measureless to man
   Down to a sunless sea.
           From Coleridge's 'Kubla Khan'

$$M = .77$$

Come into the garden, Maud,
  For the black bat, Night, has flown,
Come into the garden, Maud,
  I am here by the gate alone;
And the woodbine spices are wafted abroad,
  And the musk of the roses blown.
        From Tennyson's 'Maud'

$$M = .74$$

Take, O, take those lips away,
  That so sweetly were foresworn;
And those eyes, the break of day,
  Lights that do mislead the morn!
       From Shakespeare's song
        'Take, O, Take Those Lips Away'

$$M = .73$$

Tell me not, in mournful numbers,
  Life is but an empty dream!—
For the soul is dead that slumbers,
  And things are not what they seem.
       From Longfellow's,
       'A Psalm of Life'

$$M = .65$$

Little boy blue, come blow your horn,
The sheep's in the meadow, the cow's
  in the corn.
       From a nursery rhyme

$$M = .64$$

The white mares of the moon rush along the sky
Beating their golden hoofs upon the glass Heavens;
The white mares of the moon are all standing
    on their hind legs
Pawing at the green porcelain doors of the
    remote Heavens.
       From Amy Lowell's
       'Night Clouds'

$M = .62$

Bright Star, would I were steadfast as thou art —

From Keats' 'Last Sonnet'

$M = .57$

Hear the sledges with the bells —
Silver bells!
What a world of merriment their melody
foretells!

From Poe's 'The Bells'

$M = .51$

Onward, Christian soldiers,
  Marching as to war,
With the cross of Jesus
  Going on before.

From Baring-Gould's
'Onward Christian Soldiers'

$M = .45$

He never had much to give,
  Subscription lists knew not his name,
He was one of the many who live
  Unrecorded in charity's fame.

From E. A. Guest's
'Contribution'

It may be remarked that instances of consonantal excess appear in three of these cases, namely in the short second line of Poe's 'The Bells' ($ce = 1$), in the first line of Baring-Gould's 'Onward, Christian Soldiers' ($ce = 6$), and in the second line of Guest's 'Contribution' ($ce = 4$). As is almost obvious, the two opening lines from 'The Bells' are not among the best of Poe's in musical quality; for example, the first stanza from 'The Raven,' has an aesthetic measure M of .75. Furthermore in the free verse of Amy Lowell the 'accented' syllables were determined by ear, and the third and fourth lines were taken as 'rhyming' with the first and second respectively, because of the repetition of words.

It is well to bear in mind the precise significance claimed for these and similar results:

THE MUSICAL QUALITY IN POETRY

(1) The aesthetic measure M defined above is applicable primarily to the great body of English poetry of conventional type; the definition made can doubtless be considerably improved on the basis of further experiment.

(2) Only the musical quality of this kind of poetry is so measured. Every good poet will find it desirable to sacrifice this musical quality occasionally in order to produce some subtle musical effect or to increase expressiveness.

(3) In so far as this measure M is applied to more recent writers (such as Amy Lowell, for instance), it serves only as an indication of the presence or absence of musical quality of this conventional type.

21. Sonorous Prose

It is obviously possible to measure the musical quality of sonorous prose by the same methods. For this purpose it is only necessary to write the prose as nearly as possible in the form of verse and then to apply the same rules. For instance the following sentence from Sir Thomas Browne's 'Hydriotaphia' is so written:

> Circles and right lines limit and close all bodies,
> And the mortal right-lined-circle
> Must conclude and close up all.

As written, the sentence has an aesthetic measure M of .61 and so must be regarded as on a level with much poetry in degree of musical quality.

22. Poetry in Other Languages

As far as I have been able to make out, the aim of poetry is essentially the same, whatever the language or period. It is true that rhyme may assume different forms or may be absent as in blank verse. But the fundamental aim is always to achieve the terse, imaginative expression of a poetic idea in metric form by use of language of unusual musical quality.

23. The Rôle of Musical Quality in Poetry

In order to avoid misunderstanding I would like to emphasize once more that musical quality is only one of the essential elements in poetry, and that even this quality cannot be measured in its more delicate *nuances* by any mechanical method, such as that given above. At the same time, it seems to me that some such objective method of evaluation can play a useful if modest rôle.

Paul Valéry, the French poet and critic, has expressed the extraordinary difficulty of poetic achievement: *

> One feels clearly in the presence of a beautiful poem of some length how slight the chance is that a man could have improvised without revision, without other fatigue than that of writing or uttering what comes to his mind, an expression of thought, singularly certain, showing power in every line, harmonious throughout, and filled with ideas that are always felicitous; an expression that never fails to charm, in which there are no accidents, no marks of weakness or of lack of power, in which there are no vexatious incidents to break the enchantment and destroy the poetic universe.

Nevertheless, in achieving this complex, difficult end, the poet must take cognizance of the essential formal factors of metre and musical quality which differentiate poetry from prose. Of this necessity upon the poet, and of the others, Valéry speaks as follows:

> Behold the poet at grips with this unstable and too mixed material [of language]; constrained to speculate concerning sound and sense in turn, to achieve not only harmony and musical phrasing, but also to satisfy a variety of intellectual conditions, logic, grammar, the subject of the poem, figures and ornaments of all kinds, not to mention conventional rules. See what an effort is involved in the task of bringing to a successful end an expression of thought in which so many demands must all be miraculously satisfied!

* 'La Poésie,' Conferencia, Paris (1928), my translation.

CHAPTER IX

EARLIER AESTHETIC THEORIES

1. Introduction

IN ORDER to give a proper setting to our quantitative theory of aesthetic measure it is necessary to consider its relation to earlier aesthetic theories. The account of these developments here given can only be one of broadest outline of course, for the literature involved is extremely extensive. Our main interest will be to interpret the principal advances in terms of the quantitative theory, and also to observe how far earlier writers, beginning with Plato and Aristotle, have perceived the presence of formal elements of order in art, and what rôle they have ascribed to these elements.

The pleasurability of art as based upon its sensuous nature, its usefulness for purposes of instruction, its mystic quality due to the presence of connotative or occult formal elements of order — all these are obviously important aspects, each of which has from time to time been looked upon as of dominant importance. Thus have arisen *hedonistic*, *pedagogic*, and *mystical* theories of art and aesthetics. From the sound scientific point of view, however, it seems almost meaningless to declare that one of these aspects is the most fundamental.

In contrast with hedonistic, pedagogic, and mystical theories may be placed those *analytic* theories which attempt to discover the specific aesthetic factors involved in the several fields of art, to appraise the rôles of these factors, and then to formulate general laws so far as possible. Such theories are obviously concerned with what we have called the fundamental problem of aesthetics. Our own mathematical theory finds its place among these analytic theories, but is distinguished from the others in that it aspires to provide a quantitative solution of the fundamental problem, at least as far as the formal side of art is concerned. It seems almost obvious that aesthetics, if it is to be scientific, must be approached from the analytic point of view and must concern itself chiefly with the formal aspects of art.

For this reason we shall allude only in the briefest terms to those aesthetic writings which are not analytic in character, despite the fact that some of them are literary works of art of a very high order.

2. Plato

From very early times there are to be found numerous critical reflections concerning poetry, painting, and sculpture. In fact it is inherent in the nature of the aesthetic process that objects of the same kind provoke comparison, and that this comparison leads to increased understanding of the underlying aesthetic factors.

Now it was obvious first of all that painting and sculpture were closely akin in that they were representative or mimetic, and that poetry shared in this attribute because of its capacity to suggest a series of visual and auditory images. On the other hand poetry was seen to differ from painting and sculpture because of its much greater pedagogic power. In this respect poetry is similar to philosophy, but differs from it in that poetry conveys ideas in an indirect pleasing manner rather than in the direct neutral manner characteristic of philosophical discourse.

From his philosophic point of view Plato was led to assign an inferior position to art because of this mimetic quality: for a work of art was the imitation of an object, and these in turn were but faint copies of the fundamental Platonic Ideas; thus art, as the imitation of an imitation, could scarcely merit serious philosophic consideration. Furthermore, the admixture of the sensuous element in art did not meet with his approval. In consequence of these reflections Plato proposed to exclude poets from his ideal Republic. Such a criticism of art is evidently unsound since it ignores the fact that art is imaginative and expressive.

Plato was also interested in the problem of the beautiful, as Socrates and other philosophers had been before him. In general it may be said that for him the beautiful was not that found in art, but rather that of objects in nature. Thus in the *Hippias maior* a beautiful maiden, mare, lyre, and vase are instanced. The beauty of laws and of actions is also remarked upon.

In this dialogue several definitions of the beautiful are in turn examined and rejected: that which is fitting, or useful, or good; that which delights

EARLIER AESTHETIC THEORIES

the sight and hearing. It is evident that the first definitions are negative and therefore incomplete; they refer to those conditions which must be satisfied before beauty can be possible. The second definition is also finally rejected because no common element in the senses of sight and of hearing is found, so that beauty would appear to be two things, instead of one.

But there is an underlying common element. Sight presents to us the three-dimensional 'metric manifold' of ordinary space (in mathematical terminology); likewise, hearing presents the one-dimensional metric manifold of time. Spatial form and temporal form are therefore of the same abstract nature, and the aesthetic enjoyment of spatial and temporal objects arises in large measure from the formal relations of metric manifolds. Goethe's characterization of architecture as 'frozen music' embodies this truth suggestively.

Plato himself recognizes the importance of this mathematical element, for we read in the *Philebus*: "If arithmetic, mensuration, and weighing be taken away from any art, that which remains will not be much," and again: "For measure and proportion always pass into beauty and excellence." In connection with this second passage it is made explicitly clear that he is referring to beauty of geometric form, as exemplified for instance by a circle, and beauty of musical form, as exemplified by a pure musical note.

3. ARISTOTLE

Aristotle, inventor of formal or syllogistic logic, author of the *Poetics* and the *Rhetoric*, saw more clearly than Plato that art is expressive and not merely imitative. At the same time it was not this characteristic which seemed to him fundamental, but rather the characteristic of mathematical form: "Those are mistaken who affirm that the mathematical sciences say nothing of beauty or goodness. For they most especially discern and demonstrate the facts and definitions relating to them; for if they demonstrate the facts and definitions relating to them, though without naming the qualities in question, that is not keeping silence about them. The main elements of beauty are order, symmetry, definite limitation, and these are the chief properties that the mathematical sciences draw attention to." *

* *Metaphysics.*

In his very searching analysis of poetry, Aristotle deals mainly with aesthetic factors other than that of musical quality which we have considered. His concept of tragedy "as a representation of an action noble and complete in itself, and of appreciable magnitude, in language of special fascination, using different kinds of utterance in the different parts, given through performers and not by means of narration, and producing by pity and fear, the alleviating discharge of emotions of that nature" * recognizes this musical factor, for by 'language of special fascination' is certainly meant that involving rhythm and melody.

4. Plotinus

The Greek philosopher Plotinus is the first great representative of the mystical point of view referred to above. He refused to admit that beauty is identifiable with mere symmetry: "Beauty is rather a light that plays over the symmetry of things than the symmetry itself, and in this consists its charm. For why . . . are the more life-like statues the more beautiful, though the others be more symmetrical . . . except that this living beauty is more desirable . . . ?" † Here 'light' refers to actual light in the physical sense. According to him things are beautiful because they participate in having the attribute of rational form, characteristic of the soul, rather than because of symmetry.

The reaction of Plotinus against the point of view of Plato and Aristotle is justified to the extent that it stresses the inadequacy of any purely formal analysis of a work of art. But when he proceeds to interpret the aesthetic experience in a purely mystical manner it does not appear that any real advance is accomplished.

5. The Greek View

It appears then that the prevailing Greek view emphasized the importance of the formal elements in art. Bosanquet says (*loc. cit.*): "the one true aesthetic principle recognized by Hellenic antiquity in general" is that "beauty consists in the imaginative or sensuous expression of unity in variety. . . . The relation of whole to part — a slightly more concrete

---

* *Poetics.* See Bosanquet, *A History of Aesthetic*, London (1892).
† *Ennead.* See Bosanquet, *loc. cit.*

expression for unity in variety — has never been more perfectly elucidated or more justly appreciated than by Plato and Aristotle. . . . Moreover, the relation of the one to the many or of the part to the whole is represented in comparative purity by geometric figures, or again by rhythms or spatial intervals that bear numerical relation to one another. And for this reason Greek philosophy is inclined to select form, ratio, or proportion, as the pure and typical embodiment of beauty."

6. Luca Paciolo. Michelangelo

Although it was not until the seventeenth century that important new aesthetic ideas appeared, it is interesting to observe that even earlier mathematicians and artists were led to ascribe peculiar aesthetic merit to certain numerical proportions. Thus the mathematician Paciolo in his *De divina proportione* of 1509 considers the 'golden section' of a linear segment to be fundamentally important. This is the proportion which is found in the Golden Rectangle. Michelangelo, close friend of Paciolo, ascribed certain simple proportions to the ideal human figure.

It is obvious of course that the theory of aesthetic measure provides little justification for any mystic Pythagorean dogma in the field of aesthetics, although it recognizes the importance of simple numerical relationships in certain cases.

7. Fracastoro

What appears to be the first explicit statement that beauty must always be relative to objects of a definite class was made by the physician and poet Fracastoro in his *Naugerius, sive De poetica* of 1555. This truth is of course involved in our formulation of the fundamental aesthetic problem. It is more or less implicit in the works of Plato and Aristotle.

8. Wit and Taste

In the seventeenth and early eighteenth centuries a great deal of suggestive aesthetic discussion turned upon the discussion of 'wit' and 'taste.'*

* See B. Croce, *Aesthetic as Science of Expression* (translation by D. Ainslie), second edition, London (1922), chap. 3, for an account of this development.

It is difficult, however, to attribute the origin of these notions with certainty to particular writers.

Wit is taken as synonymous with genius and creative imagination, and as standing in contrast with pure intellect. Similarly taste refers to the intuitive aesthetic judgment based on aesthetic feeling, which may, however, be in part susceptible of intellectual analysis.

The notion of taste is inherent in our analysis of the act of aesthetic perception: first the effort of attention, then the intuitive aesthetic judgment, dependent upon taste, and finally analysis.

Wit or the faculty of creative imagination is evidently closely allied with taste or the appreciative faculty. In fact wit may be regarded as taste transposed to a higher key. It is in this light that we consider it briefly in the following chapter.

9. Descartes

The rationalistic philosophy of Descartes gave to reasoning a position of first importance: *cogito ergo sum*. Himself a creative mathematician, he regarded mathematical reasoning as the model from which to start. The formalistic universe to which a mixture of speculation and reason led him contained many metaphysical, physical, biological, physiological, and psychological doctrines of great originality and interest. Unfortunately, in his dualistic account of mind and matter, imagination was regarded as caused by the play of the animal spirits upon the mind. In consequence, poetry and other works of art were tolerated only in so far as they were in accord with reason. It is very significant of this general attitude that Descartes scarcely refers to aesthetic questions in his extensive writings.

10. Leibnitz

The formalistic universe of Leibnitz, also a great mathematician, provided for every conceivable type of being, each having its representative monad. Thus he was able to admit aesthetic facts without any difficulty, since aesthetic perceptions and judgments were valid forms of knowledge. Such knowledge is as *clear* as intellectual knowledge, but differs from the latter in that it is *confused* instead of *distinct*.* Here he follows a classification employed earlier by Descartes.

* *De cognitione, veritate et ideis* (1684).

EARLIER AESTHETIC THEORIES

Leibnitz's celebrated enigmatic definition of music as "counting performed by the mind without knowing that it is counting" will be seen to be consonant with our thesis that the number of certain orderly relations among the notes, estimated intuitively, measures the aesthetic effect. This definition of Leibnitz may also be regarded as partly substantiated later by certain musical researches to which we shall refer.

11. BOILEAU. DE CROUSAZ

Boileau and other writers of the Cartesian school soon tried to extend Cartesian doctrines to the aesthetic realm. This attempt led inevitably to a completely intellectualized point of view. De Crousaz says in his *Traité du beau* of 1724: "Good taste makes us appreciate at first by feeling that which reason would have approved. . . ." But this claim is certainly exaggerated in all of those cases where important connotative elements enter, as we have frequently pointed out.

The main factors of the beautiful were considered by de Crousaz to be variety, unity, regularity, order, and proportion. Evidently these are essentially the same mathematical elements which were specified by Plato and Aristotle.

12. VICO

Vico, a jurist known generally for his 'philosophy of history,' in common with many others of his period reacted strongly against the intellectualistic view of poetic art. He considered imagination as embodied in poetry to be of equal importance with intellect as embodied in metaphysics, and independent of it. Furthermore he identified poetry with language, thus emphasizing first of all the expressive power of art.* For this reason Croce, himself of the same school, regards Vico as "the real revolutionary who by putting aside the concept of probability and conceiving imagination in a novel manner actually discovered the true nature of poetry and art and, so to speak, invented the science of Aesthetic. . . ." †

However, after the acceptance of such a vigorous affirmation of the expressiveness of art, it remains (in our opinion) to discover how art is

* *Scienza nuova prima* (1725).
† *Loc. cit.*

made expressive. This analytic phase was repellent to Vico. To us on the contrary it appears to be the chief part of the science of aesthetics.

13. Rameau

The musician Rameau in his *Traité de l'harmonie* of 1722 penetrated deeply into the nature of harmony and thus contributed to the understanding of certain aesthetic effects in music. He observed that a musical note is in general composite, being composed of a pure fundamental note and overtones which can be heard, and that notes differing by an octave are so similar in their aesthetic effect as to be almost identical. These facts lead directly, as we have seen, to the understanding of the Western scale. They also lead to the notion of the fundamental bass or root of a chord, due to Rameau, and explain why this bass must in general proceed by a fourth or fifth, up or down, to its harmonically nearest notes.

The cogency of his development seemed to Rameau so complete that he entitled a later work (1750) *Démonstration du principe de l'harmonie*. However, the acute mathematician d'Alembert gave a clear presentation of Rameau's work in which he stated the requisite empirical rules without attempt at their demonstration.*

Rameau's treatment of chords and of chordal sequences, and other later treatments of harmony, differ from our own in two important respects: they are qualitative rather than quantitative since they aim mainly to exclude inadmissible chords and chordal sequences; empirical exceptions are introduced wherever necessary, whereas our theory proceeds uniformly on the basis of the specific evaluation of simple aesthetic factors.

14. Euler

In his *Tentamen novae theoriae Musicae* of 1739, the mathematician Euler developed a theory of consonance based upon the Pythagorean law. This is interpreted in the general sense that the smaller the integers expressing the vibration ratios, the more consonant the notes involved will be. In this way he is led to a simple empirical rule for estimating the degree of consonance or harmony of any interval or chord, which in general accords with the observed facts.

* *Éléments de musique, suivant les principes de M. Rameau* (1762).

EARLIER AESTHETIC THEORIES

The degree of consonance is of course entirely distinct from that of agreeableness or aesthetic measure. For instance, unison and the octave are the most harmonious of all intervals, but are not the most agreeable. Nevertheless, it is extremely interesting that Euler should have formulated a quantitative rule for the measurement of consonance.

Euler's general concept of the nature of aesthetic enjoyment was in entire agreement with our own as may be gathered from the following general account of it as given by Helmholtz: * "The more easily we perceive the order which characterizes the objects contemplated, the more simple and more perfect will they appear, and the more easily and joyfully shall we acknowledge them. But an order which costs trouble to discover, although it will indeed also please us, will associate with that pleasure a certain degree of weariness and sadness."

15. HOGARTH

The artist Hogarth in his *Analysis of Beauty* of 1753 attempted an analysis of the aesthetic factors involved in painting. As we shall point out in Chapter X, the aesthetic problem here is vague and difficult. He did no more than enumerate formal factors such as symmetry, variety, uniformity, simplicity, intricacy, quantity, and convincing representation. He ascribed an especial beauty to a serpentine line which he called the "Line of Beauty."

16. BURKE

At about the same time the statesman and philosopher Edmund Burke in his *Philosophical Enquiry into the Origin of our Ideas on the Sublime and the Beautiful* of 1756 separated the sensuous and imaginative factors in the work of art. These latter are conceived of as essentially mimetic. Furthermore he endeavored to classify the various aesthetic factors upon which depends the beauty of an object.

17. HEMSTERHUIS

The philosopher Hemsterhuis in a 'Lettre sur la sculpture' published in 1769 gave a definition of the beautiful which has become very well known: "the beautiful is that which gives the greatest number of ideas in the shortest space of time."

---

* *Tonempfindungen* (1862), also *Sensations of Tone* (translated by A. J. Ellis), fourth edition, London (1912).

AESTHETIC MEASURE

This definition contains much of the essence of our fundamental formula taken in a qualitative sense. The ideas to which he refers correspond to the connotative elements of order in the aesthetic object.

18. Kant

The great philosopher Kant devoted much of his attention to aesthetic questions. To him aesthetical ideas, expressed in works of art, are supplementary to logical ideas or concepts and reinforce them: such works "make us think more than we can express in a given concept by means of words and give us an aesthetic idea which serves to this rational idea instead of a logical representation."* Evidently this doctrine emphasizes the expressive character of art, and points toward the associations and feelings aroused by the contemplation of a work of art as of vital importance.

Kant brought out clearly the distinction between sensuous, emotional, moral, or intellectual feeling, and aesthetic feeling, to which we alluded at the beginning: sensuous or emotional feeling is excluded because the beautiful must please "without interest"; moral feeling is excluded because it must please "without the representation of an end"; intellectual feeling, because it must please "without concepts." Throughout Kant's writings there is evident a strong tendency to adopt a mystical view towards art. There is little which can be regarded as analytical.

19. Schiller. Hegel

The followers of Kant and his metaphysical idealism continued in the same speculative realm of thought. Their aesthetic writings culminate in striking mystic phrases such as: beauty is "living form" (Schiller); art "cancels matter through form" (Schiller); beauty is the "sensible appearance of the Idea" (Hegel).

20. Herbart

Among those of the Kantian school, Herbart seems to us the most suggestive in his point of view towards aesthetics, despite a certain dry formalism.

* *Kritik der Urtheilskraft* (1790).

EARLIER AESTHETIC THEORIES

For Herbart, beauty consists primarily in relations, that is, in elements of order, according to our terminology. Furthermore art is two-sided: it possesses 'content,' which is not properly aesthetic, and 'form,' which is of its essence.* Evidently his division into form and content corresponds closely with our own into the formal and connotative sides.

21. Schleiermacher

The theologian and philosopher Schleiermacher insisted upon the inspired expressional nature of art, as so many of the Kantian school and others had done. What especially interests us is his realization that a work of art must be compared with others of the same kind. In his insistence upon this fact he is very explicit: "There is no difference in works of art except in so far as they can be compared in respect of artistic perfection." "In this respect the biggest, most complicated canvas is on a level with the smallest arabesque, the longest poem with the shortest." "This proposition must be adhered to absolutely, if irrelevant elements are not to enter everywhere." †

By this insistence Schleiermacher clarified the fundamental problem of aesthetics. In the various applications of our quantitative theory we have seen how essential it is that the class of aesthetic objects be closely prescribed. Hence our conclusions are entirely in agreement with Schleiermacher's in this respect.

22. Poe

The American poet Poe, always a theorist and many times an extravagant one, was led by his study of poetry to an aesthetic theory in which is formulated, apparently for the first time, the conjecture that aesthetic elements of order have a definite weight. The following quotation from the 'Rationale of Verse' of 1843 embodies this conjecture:

Let us examine a crystal. We are at once interested by an equality between the sides and between the angles of one of its faces: the equality of the sides pleases us; that of the angles doubles the pleasure. On bringing to view a second face in all respects similar to the first, this pleasure seems to be squared; on

---

* *Einleitung in die Philosophie* (1813).
† *Vorlesungen über Ästhetik* (only published in 1842 after his death).

bringing to view a third it appears to be cubed, and so on. I have no doubt, indeed, that the delight experienced, if measurable, would be found to have exact mathematical relations such as I suggest; that is to say, as far as a certain point, beyond which there would be a decrease in similar relations.

So far as I know, this is the only affirmation of the kind to be found in earlier aesthetic theories.

In the application of this general idea to poetry Poe dealt with the formal elements of rhyme (interpreted to include alliteration and assonance) and metre. For rhyme and metre were respectively the 'equality' of sound and the 'equality' of time, both appreciated by the ear. He formulated no precise measure of the musical quality in poetry (see section 8, Chapter VIII), but his poems are very remarkable in this respect.

23. SPENCER

The positivist philosopher Spencer made several penetrating studies in the field of aesthetics. In his *Philosophy of Style* of 1852 he asserts that the effective cause of style is economy of effort. The notion of economy of effort, independently developed later by Fechner, is evidently in accord with our theory. In the nearly contemporaneous *Origins of Architectural Styles* he ascribes the beauty of architectural form to uniformity and symmetry, qualities exhibited in the forms of nature: in other words the aesthetic effectiveness of symmetry in architecture, for instance, is due in part to its association with the symmetry of the human body and other natural forms. Moreover, in his *Origin and Function of Music* (1857) he advances the view that music is derivative from language. We have observed that the connotative element in music, produced by this linguistic origin, is beyond the reach of any analytic theory such as that here advanced.

24. HELMHOLTZ

Helmholtz, physiologist, physicist, and mathematician, undertook the systematic examination of the physical and physiological basis of sensations of tone. The *Tonempfindungen* of 1862, embodying his results, constitutes the veritable *Principia* of the subject. In particular the existence of summation and difference tones is established and explained upon a mathematical basis.

His genetic account (following Rameau) of chords and the Western diatonic scale is complete in that these are shown to arise naturally. In other words the associative structure involved is traced in all of its ramifications. Helmholtz only considers briefly the complicated question of chordal sequences, and in this respect does not go much beyond Rameau.

In his work there is to be found no indication of the quantitative outlook. However, he clearly adheres to the general analytic point of view, for he says: "No doubt is now entertained that beauty is subject to laws and rules dependent on the nature of human intelligence" which "are not consciously present to the mind, either of the artist . . . or the observer. . . ." Indeed "it is an essential condition that the whole extent of the regularity and design of a work of art should not be apprehended consciously. It is precisely from that part of its regular subjection to reason which escapes our conscious apprehension that a work of art exalts and delights us." Thus Helmholtz believes that art depends upon definite laws which may be discovered. He denies, however, that art can satisfy after its structural laws are understood.

It seems to us that this last conclusion is not justified. In fact it is in an unusual density of elements of order, obvious, and more or less concealed, that we have found the secret of successful musical form. These types of order are so varied and numerous that the same occult aesthetic effect is felt regardless of the possibility of a systematic enumeration revealing the constituent elements.

25. Sylvester

The mathematician and casual poet Sylvester undertook in his *Laws of Verse* published in 1870 to reduce versification to definite principles. It is evident that he was influenced by the earlier work of Poe in this direction, with whom he finds himself in general agreement. Although Sylvester dealt but little with metre, his contribution is to be regarded as a distinct advance in the aesthetics of verse.

Sylvester divides the technical (formal) side of verse, which he calls "Rhythmic," into three branches: "Metric," about which he accepts Poe's doctrines; "Chromatic," dealing with the tonal side into which he does not enter; and "Synectic," concerned with the "continuous aspect of the Art." In Synectic, the central concept is that of "Phonetic Syzygy," "to which we must attend in order to secure that coherence, compactness, and ring of true metal, without which no versification deserves the name of poetry."

This concept of Phonetic Syzygy was seen in Chapter VIII (section 9) to correspond qualitatively to the aesthetic measure of musical quality, although it was not thought of quantitatively by Sylvester.

26. Hanslick

The philosopher Hanslick has exerted a strong influence against the opinion that music is beautiful primarily because of its mimetic linguistic power rather than as the embodiment of abstract form.*

In speaking of the beautiful in music he says the following:

> *Its nature is specifically musical.* By this we mean that its beauty is not contingent upon, or in need of any subject introduced from without, but that it consists wholly of sounds artistically combined. The ingenious co-ordination of intrinsically pleasing sounds, their consonance and contrast, their flight and reapproach, their increasing and diminishing strength — this it is, which in free and unimpeded forms presents itself to our mental vision.
>
> It is extremely difficult to define this self-subsistent and specifically musical beauty.
>
> A 'philosophical foundation of music' would first of all require us, then, to determine the definite conceptions which are invariably connected with each musical element and the nature of this connection, . . . a most arduous though not an impossible task.
>
> There is no art which, like music, uses up so quickly such a variety of forms.
>
> Mathematics, though furnishing an indispensable key to the study of the physical aspect of music, must not be overrated. . . . No mathematical calculation ever extends into a composition. . . .

Evidently Hanslick goes beyond Helmholtz in affirming that musical beauty is formal and is made up of elements of order which are complex but can be effectively analyzed. He perceives also that each type of music is limited by the allowance of means employed, as we have suggested.

His reference to mathematics is evidently to its use in physical acoustics. Our theory is of course devoid of any mathematics except that of mere enumeration; moreover we do not suppose that the elements of order are counted when music is heard, but rather that they are intuitively felt in their aggregate weight.

* *Vom Musikalisch-Schönen* (1874), also *The Beautiful in Music* (translation by G. Cohen, 1891).

EARLIER AESTHETIC THEORIES

27. Fechner

With the rise of modern experimental psychology it was inevitable that the field of aesthetics should be approached in the light of the new ideas which it made available. The well known psychologist Fechner attempted to found a science of aesthetics "from below" of this kind.* In the long list of 'principles' and 'laws' to which he is led, an important rôle is assigned to association; we have seen how important this rôle is. Fechner contrasts the direct factor in aesthetic perception with this associative factor. For example, in the perception of an orange the direct factor would be essentially that of a yellow sphere; the indirect, of a delicious tropical fruit, etc.

Undoubtedly Fechner did not consider the spherical symmetry of the orange as giving rise to a specific association. Our reason for so regarding it may be stated as follows: all objects possessing such symmetry are associated by means of the uniform tactual and visual technique involved; in general this symmetry is desirable, so that the association has a positive tone of feeling.

Aside from his recognition of the importance of associations for aesthetic perception, Fechner's main contribution was methodological. He was the first to treat the fundamental aesthetic problem for simple classes of aesthetic objects by direct experiment. However, as we have already indicated, in his treatment of rectangular form the actual experimental results were inconclusive.

28. Lanier

The American poet Lanier wrote in 1880 an important book, *The Science of English Verse*, of which the main point is the definite parallelism between poetry and music: "Perhaps no one will find difficulty in accepting the assertion that when formal poetry or verse . . . is repeated aloud, it impresses itself upon the ear as verse only by means of certain relations existing among its component words considered purely as sounds, without reference to their associated ideas." Furthermore he undertakes to establish that "the sound-relations which constitute music are the same with those which constitute verse, and that the main distinction between music and verse is,

* *Vorschule der Aesthetik* (1876).

when stated with scientific precision, the difference between the scale of tones used in music and the scale of tones used by the human speaking-voice." In developing this thesis, Lanier confines his attention mainly to the phenomenon of rhythm, where his views differ substantially from those of Poe. There is of course no effort at a quantitative consideration on the part of Lanier.

29. Lipps

The psychologist Lipps stressed the importance of the empathetic factor in aesthetic perception, by which the self is identified with the artistic object.* Such identification is obviously important in the appreciation of a statue and in similar cases. Here the complete set of associations induced by the act of attentive perception necessarily leads to such an empathetic response.

30. Gurney

The philosopher and psychologist Gurney in his *Power of Sound* of 1889 undertook to appraise the rôle of the formal elements of order in music. Unfortunately he overlooks the presence of any but the most obvious formal elements, and does not perceive in the least how such order should be measured. In consequence he is forced to the conclusion that music is "Ideal Motion" and that "the essential characteristic of the complete Ideal Motion is an absolutely unique beauty perceived by an absolutely unique faculty. . . ." Of course such a conclusion is entirely mystical. A specific solution of the problem of melody, which Gurney thus declares to be insoluble, has been tentatively proposed by us in an earlier chapter.

31. Croce

The aesthetician and philosopher Croce follows in the steps of Vico and others by insisting that art is expressive: Art is the "expression of impressions . . ."; † art is "lyrical intuition." ‡ For him knowledge is divided into intuitive knowledge and conceptual knowledge; the first finds its expression in art, the second in science and philosophy.

* *Aesthetische Faktoren der Raumanschauung* (1891).
† *Loc. cit.* His *Estetica* appeared first in 1902.
‡ *Aesthetics*, Encyclopedia Britannica, 14th edition (1929).

EARLIER AESTHETIC THEORIES

Such general philosophical definitions and classifications, however true, can never serve as the point of departure for a science of aesthetics. They are self-limited and form a kind of philosophic citadel from which an attack upon any and all more definite conclusions can be conveniently made.

32. Ross. Pope

My colleague Dr. Denman W. Ross has systematically treated design and painting by analysis of the formal aesthetic factors which enter. In his *Theory of Pure Design* of 1907 he says:

> The Beautiful is revealed, always, so far as I know, in the forms of Order, in the modes of Harmony, of Balance, or of Rhythm. While there are many instances of Harmony, Balance, and Rhythm which are not particularly beautiful, there is, I believe, nothing really beautiful which is not orderly in one or the other, in two, or in all three of these modes. In seeking the Beautiful, therefore, we look for it in instances of Order, in instances of Harmony, Balance, and Rhythm. We shall find it in what may be called supreme instances.

Thus Ross defines the beautiful to be a "supreme instance of Order." This concept, although entirely qualitative, is evidently akin to that embodied in our general theory.

Ross has classified the formal elements in design and painting as due to repetition, balance, and sequence as applied to tones, sizes, and shapes.* However, the conditions under which such elements become "emotionally, as opposed to, intellectually, effective" are none too apparent; the importance of this question has thus been pointed out by my colleague Professor Arthur Pope.†

33. The Eastern View of Art

So far as I can discover, the general analytic treatment of art, such as leads to aesthetics properly so called, is not to be found in the East. Instead there are charming literary anecdotes about artists and their work, as well as definite technical rules for the artist.

There are some slight indications of a more general point of view. For instance in Chinese art there are such pronouncements as the Six Canons of Painting of Hsieh Ho: (1) vitality, (2) anatomical structure, (3) con-

---

* *On Drawing and Painting* (1912).
† *An Introduction to the Language of Drawing and Painting:* I, *The Painter's Terms* (1929).

formity with nature, (4) suitability of coloring, (5) artistic composition, (6) finish.* Likewise in Indian music there is considerable elaboration of many modes of musical scales.

34. Concluding Remarks

These appear to us to be some of the main advances which have been made in the domain of analytic aesthetics.

In this brief account no attention has been given various subjective theories such as are often formulated by poets and other creative artists because of their pragmatic usefulness. It is to be expected that these have many points in common with theories of more objective type. Of course the artist, on account of his highly developed powers of intuitive judgment, is not apt to stray far from the right track because of an inaccurate theory. In fact such a theory may prove valuable by suggesting novel combinations and experiments.

In our opinion the above account of previous developments serves to bring out clearly the fact that the principal earlier advances in analytic aesthetics can be conveniently and adequately interpreted in terms of the theory of aesthetic measure, at least in so far as they are concerned with the precise formal side rather than the elusive connotative side of the aesthetic experience.

* See H. A. Giles, *History of Chinese Pictorial Art*, London (1905).

CHAPTER X

ART AND AESTHETICS

1. Types of Aesthetic Experience. Art

A LARGE part of ordinary experience admits of interpretation from the aesthetic point of view. In fact consciousness points successively to objects in experience which are identified by direct association as belonging to certain classes. Whenever such an object can be isolated to the extent that another object of the same class may be substituted in its place, it is generally found that certain objects are preferred to others. Thus the objects of the class fall into an orderly arrangement with regard to 'desirability' in the given context. If some aesthetic value is attributed to this desirability, the objects become aesthetic objects to that degree, and the feeling of desirability is called aesthetic feeling. In this sense the types of aesthetic fields are almost as numerous as the generic names of objects having visual or auditory reference.

Moreover our previous limitation of aesthetic perceptions to the auditory and visual domains was merely made for practical reasons and eliminated altogether too much; for instance, a system of laws may be beautiful, or a mathematical proof may be elegant, although no auditory or visual experience is directly involved in either case. It would seem indeed that all feeling of desirability which is more than mere appetite has some claim to be regarded as aesthetic feeling.

The domain of art, although of similar vast extent, refers more especially to those aesthetic fields in which the objects admit of being freely created by an artist.

2. The Variability of Aesthetic Values

Owing to constantly changing external conditions as well as the ardent human desire for novelty, aesthetic values are extremely variable. It is only in technical fields of art, where works of art are preserved, and the past is compared with the present, that this change is slow and evolutionary in character.

Moreover there are very special simple cases in which aesthetic values have taken on an almost crystalline structure. For example the appreciation of polygonal forms seems to be established in a definite way. It has been precisely to cases of this sort that we have applied the theory of aesthetic measure.

3. Qualitative and Quantitative Aesthetics

The theory of aesthetic measure cannot be expected to be more precise than the fluidic aesthetic values with which it is concerned. This does not mean, of course, that the general theory will not apply whenever definite standards of aesthetic appreciation have been set up.

In this connection it is instructive to recall the physiological interpretation given at the outset. According to it an aesthetic field corresponds to a network of associative nerve fibres in the brain. From this point of view, it is only after the normal trend of these associations has been established through considerable comparative aesthetic experience, that the network takes a characteristic form.

Qualitative aesthetics may be defined to be that part of aesthetics which traces the general nature of the associative network by furnishing a rough description of the aesthetic factors.

On the other hand, quantitative aesthetics, based upon the theory of aesthetic measure, must limit itself to those cases where the associative network has attained a characteristic development, so that not only the factors may be isolated but their relative importance may be estimated. In this case the network is of a much more definite and formal type.

4. The Qualitative Application of the Aesthetic Formula

Despite the limitation to which the quantitative application of the aesthetic formula is thus subjected, it is still possible to conceive of it as *qualitatively* applicable in more general cases, since the theory upon which it is based is an entirely general one.

In such qualitative application the effective elements of order and the approximate complexity have to be determined. However, indices are not definitely assigned, but are roughly estimated. In other words what Pope has called the 'emotionally appreciable' factors must be separated from those of no substantial effect, and their relative importance assessed, by means of the aesthetic judgment.

This kind of use of the formula leads at once to certain well known aesthetic maxims:

(1) Unify as far as possible without loss of variety (that is, diminish the complexity C without decrease of the order O).

(2) Achieve variety in so far as possible without loss of unity (that is, increase O without increase of C).

(3) This 'unity in variety' must be found in the several parts as well as in the whole (that is, the order and complexity of the parts enter into the order and complexity of the whole).

From our standpoint such maxims are all contained implicitly in the aesthetic formula $M = O/C$.

5. Decorative Design

We propose to indicate very briefly how such qualitative application of the aesthetic formula may be made in certain fields of art. Among these fields decorative design is the simplest, since the connotative elements of order are restricted to purely conventional floral or other representation.

In decorative art there is always a first requirement that the design be fitting to the end in view. We may regard this utilitarian requirement as one involving negative elements of order. Aside from this primary general requirement, a decorative design in a plane surface may be treated as an ornamental pattern (rectilinear, curvilinear, or mixtilinear), that is, as a hierarchical aggregate of ordinary ornaments. The aesthetic measure of the complete design is to be conceived as a kind of average of the aesthetic measures of the constituent ornaments, as in the case of tilings (see Chapter III).

With regard to the positive elements of order O in the constituent ornaments, we are led to list those already considered: vertical symmetry; balance; rotational symmetry; relation to a rectangular network of lines or to other one- or two-dimensional ornaments; repetition; similarity; spiral form; circular form; floral form. Furthermore we may include even more elaborate types of "Repetition, Sequence, and Balance" as indicated by Ross.

AESTHETIC MEASURE

Likewise we are led to list certain negative elements of order such as the following: ambiguity of form; lines or curves leading into a design and not continued into any straight line or curve; more than two types of directions or one type of niche in any constituent polygonal part of the design (see Chapter II); lack of symmetry; lack of suitable centers of interest.

We shall not attempt to refer to the analogous evaluation of three-dimensional decorative design, except to observe that here the elements of order are either the same as those in two-dimensional design, or are natural extensions thereof.

6. Painting, Sculpture, and Architecture

The art of painting is of particular interest. It involves not only the formal elements of order inherent in decorative design, but also those based upon color; moreover, by means of representation, it can call upon numberless associations corresponding to connotative elements of order. Great painting has developed typically in the East and in the West, but these types are sufficiently alike so that their qualities are generally appreciated.

The 'complexity' of paintings is usually so considerable that they are analogous to ornamental patterns whose constituent ornaments must be appreciated one by one. However, it is decidedly interesting to remark in this connection how a fine composition is always arranged so as to be easily comprehensible. The factors involved in such a composition are so varied and subtle that only a partial analysis is possible.

By way of illustration I give an obvious linear analysis of Correggio's 'Danae' in Plate XXII opposite, which shows some of the most important divisions and directions of the canvas; and at my request Professor Pope has kindly indicated his analysis of Veronese's 'Family of Darius Before Alexander' (shown in the same Plate):

This painting by Veronese illustrates the simplicity in the general arrangement of principal masses within the rectangle of the enclosing enframement that is constantly found in the works of the Venetian masters of the sixteenth century. A diagonal drawn from the upper right to the lower left corner divides the main figure part of the composition from the architectural background part. The stronger intensities of color and what are on the whole darker tones come in the lower right-hand triangle, with lighter tones and lesser contrasts in the

PLATE XXII

Correggio's 'Danae'

Braun photograph

Veronese's 'Family of Darius before Alexander'

Alinari photograph

upper left. Where dark is broken up into the upper triangle, as in the dark of the central male figure, compensating light is broken into the lower, as in the patch of ermine on the shoulder of the central female figure. A similar interchange of light and dark occurs in the dark mass of servants at the left breaking into the lighter main areas and the light on the dog and small boy breaking into the dark area at the right. There are many other similar interchanges. As is usual in Venetian painting, the central axis of the composition is strongly accented by the placing of the principal group of figures and by the pylon in back of them. The architectural background produces a network of vertical and horizontal lines against which the figures are placed. There is a very subtle and complex organization in the formal arrangement of the figures, and in the dramatic action, but all this is bound together within the general enframement in the simplest possible way.

Evidently the aesthetic factors here specified by Pope are largely of the formal type which we have considered.

In painting we may with approximate accuracy separate the elements of order into those of form and of color, and consider the two separately. In other words, we may consider a painting with respect to its composition and its color effects.

As far as the composition is concerned, these elements of order are of the following three types: (1) the formal elements of order inherent in two-dimensional design; (2) the kindred formal elements of order inherent in the three-dimensional objects represented; (3) the connotative elements of order which arise from the representation.

With reference to (1) it is to be noted that there should be a natural primary center of interest in the painting and also suitable secondary centers. Such a primary center of interest is often taken in the central vertical line of the painting or at least near to it. The elements (2) are of course taken to be the same as in the three-dimensional objects represented. Thus if a sphere be represented, the elements of order inherent in the sphere would naturally be considered to be present. Finally there are the connotative elements (3) which play a decisive part; a good painting requires a suitable subject just as much as a poem requires a poetical idea.

The color spectrum, upon which the elements of order involving color necessarily depend, is not without interesting analogy to the gamut of musical tones. The simpler the palette is, the less will be the complexity, so that the palette should be as restricted as the subject permits. Evidently the eye appreciates the repetition of a color, a graded sequence of colors, and a balance of colors or of light and dark values about the centers of interest, in accordance with the classification of Ross. All of these elements of order are of definite aesthetic importance.

In sculpture and architecture a similar analysis is obviously possible. But in the latter field there is no representation, while the requirement of functional fitness becomes a connotative factor of primary importance. The photograph of the Taj Mahal shown in Plate XXIII illustrates a number of the formal elements already mentioned: in particular, vertical symmetry, balance, relation to a rectangular network, circular form, and especially the multiple repetition of the characteristic arch-dome shape, which gives the fundamental motive.

7. Western Music

As has been remarked, actual Western music can be considered as involving a composite of melodic, harmonic, and rhythmic elements of order. In this sense the possibility of a qualitative application of the basic formula is obvious. Of course in ordinary musical compositions there will enter more elaborate forms than those which we have considered, although probably these have analogues in the simple cases which we have discussed. Connotative factors also enter, due in large part to the similarity of music and emotional utterance.

8. Eastern Music

In the East there has been developed a great variety of highly interesting types of music. Although in most cases the particular scale selected is not the Western scale, nevertheless it is usually near enough to the diatonic or pentatonic scale so that Eastern music can be translated into a nearly equivalent Western form. To be sure, the claim of a genuine quarter tone scale is made in India and elsewhere; but it is hard to see how such quarter tones are to be regarded otherwise than as flat or sharp forms of ordinary harmonics of a fundamental note. The frequent occurrence of perfect fourths, perfect fifths, and octaves in Eastern music shows that such harmonic intervals are inevitable.

Harmony is almost lacking in Eastern music; but grace notes and drone notes are often used, and of course fulfill a substantially analogous rôle.

PLATE XXIII

South Façade of Taj Mahal, Agra

Johnston and Hoffmann photograph

On the other hand the use of rhythm in music is perhaps more highly developed in the East than in the West. As for melody, what is usually desired in Eastern music is a very simple theme with the proper texture. It is to be remembered that pure music is much less common in the East, where music is generally employed in connection with singing, recitative, or dancing.

Notwithstanding these differences and the diversity of scales employed, the Western ear can enjoy Eastern music readily in many of its forms, and the reasons are easily understood: rhythm is the same as among us; fixed scales are employed, usually containing certain diatonic points of reference and simple harmonic intervals; there is much use of repetition and of melodic sequence as in our own music; and, finally, there is often a kind of tonic note. It does not seem to me that the diverse types of Eastern music are likely to admit the application of the aesthetic formula in the same way as does the music of the West. I should expect such an application to be feasible only in respect to certain simple elements of order, comparable with those of alliteration and assonance in poetry, upon which the characteristic quality of each type depends.

9. Evolution in Art

Art, taken in a wide sense, is expressive of the vast universe of intuitive impressions which surrounds and sustains the central nucleus of conceptual knowledge. There is no indication whatsoever that the central nucleus will absorb the whole, as Hegel believed, nor that the growth of the whole will not continue indefinitely.

In consequence, indefinite further evolution of art is certain. But this favorable evolution is at first sight only assured through the appearance of new values on the connotative side of art. Will not the formal side of art sink finally into relative insignificance? This is the side to which we have devoted attention as being the most characteristic. It seems to me that this pessimistic conclusion is not justified, and that the formal and connotative sides will continue to advance *pari passu*. In fact I believe that ideal aesthetic expression must always unite formal and connotative elements of order to an almost unbelievable degree.

AESTHETIC MEASURE

10. Creative Art

Any satisfactory account of aesthetics must take cognizance of the creative side of art. It has been stated earlier (Chapter IX) that creative power is the aesthetic judgment transposed to a higher key.* This assertion must be justified.

It is obvious that the highly developed aesthetic judgment acts intuitively and rapidly. If there be superadded the mental energy and interest requisite for aesthetic experimentation, certain of the results will be adjudged more satisfactory than others, and will be improved, and may lead finally to works of art. Such experiments can not be wholly arbitrary but must be well directed.

Now it seems to me that the postulation of genius in any mystical sense is unnecessary. For, under the conditions specified, the experimentation will proceed automatically in an advantageous direction just because the aesthetic judgment will find mere repetition to be dull and will be attracted by interesting modifications.

In this process the explicit analysis of aesthetic factors and the formulation of aesthetic theories, however rudimentary, are inevitable and are of fundamental importance to the creative artist. Indeed, as was observed at the outset, the act of aesthetic perception is made up of three phases: attentive perception; aesthetic enjoyment; and the explicit recognition of order or harmony in the object. Thus the analytic phase appears as an inevitable part of aesthetic experience. The more extensive this experience is, the more definite becomes the analysis.

It is evident that accurate and intelligent analysis will in general prove the best guide to creative experiment. Occasionally it may happen that an apparently crude analysis may suggest unexplored possibilities that have been completely overlooked. But, in general, when the theories of an artist overbalance his aesthetic judgment and experience, he is likely to produce what may be termed 'puzzle-art.' This kind of art has been exemplified in many experiments of recent decades. Any novel artistic form which cannot be appreciated without advance knowledge of the theory underlying it may be suspected of falling into the category of puzzle-art.

---

* Lowes reaches a like conclusion in his notable work, *The Road to Xanadu*, Boston (1927).

ART AND AESTHETICS

In the inevitable analytic accompaniment of the creative process, the theory of aesthetic measure is capable of performing a double service: it gives a simple, unified account of the aesthetic experience; and it provides means for the systematic analysis of typical aesthetic fields, yielding not only a *catalogue raisonné* of the aesthetic factors involved but also an estimate of their relative importance.

The future of art is unpredictable, and on that account all the more alluring. Who can say what new kinds of associative networks possessing aesthetic significance are yet to be realized and elaborated? All these will certainly lead to further modes of artistic expression.

As always, the highest creative art will follow new paths unerringly because it is in command of the resources of the past. In this connection Croce's dictum that art is the 'expression of impressions' may be paraphrased to the effect that the most vital art is the expression in definitive form of new and significant impressions.

INDEX

INDEX

Aesthetic classes, 3, 11, 195, 201, 209
Aesthetic experience, 3, 4, 14, 209
Aesthetic factors, viii, 19, 210, 211
  in art, ix, 191, 216, 217
  in music, 88
  in ornaments, 57, 63, 64
  in poetry, 170–175
  in polygonal form, 19–33
  in vase form, 67, 85
Aesthetic feeling, 3, 6, 200, 209
Aesthetic formula, vii, ix, 3, 4, 11–15, chap. 1
  and aesthetic maxims, 211
  application of, viii, 13, 14, 210–215
  as *deus ex machina*, ix
  diagram of, 14
  for chordal sequences, 128, 129, 135–145, in particular 143–145
  for chords, 108–113
  for melody, 160–164
  for musical quality in poetry, 178–184
  for polygons, 33–44, in particular 42–44
  for simple rectilinear ornaments, 57–60
  for tilings, 64, 65
  for vases, 76–78
  justification by analogy of, 11, 12
  mathematical justification of, 12, 13
  psychological basis of, 3–12
Aesthetic judgment, 3, 11, 17, 196, 216
Aesthetic maxims, 211
Aesthetic measure, viii, ix, 3, 4, 11–15, 197, 198, 199, 201, 202
  in art, 211–215
  of Beethoven chorale, 156–160
  of 8 Chinese vases, 78–84
  of 144 chordal sequences, 146–151
  of chords, 113–127
  of 3 experimental melodies, 164–168
  of experimental poem, 185, 186
  of 3 experimental vases, 84, 85
  of 10 poetic selections, 186–188
  of 90 polygons, 44–46
  of 30 simple rectilinear ornaments, 62, 63
  of sonorous prose, 189
  of 12 tilings, 66
  of verses of 'Kubla Khan,' 184, 185
  origin of theory of, vii
Aesthetic perception, 3–5
Aesthetic problem, 3, 4, 11
  of harmony, 89, 97, 98, 128
  of melody, 152, 153
  of musical form, 87–89
  of ornaments, 56
  of polygonal form, 16, 17
  of rhythm, 169
  of vase form, 67, 68
Aesthetic theories, chap. 9
Aesthetic value, 3, 209, 210
Aesthetics, 3
  extent of, 209
  qualitative and quantitative, 210
  theories of, 216, chap. 9
Ainslie, D., 195
Alembert, J. L. R. d', 198
Alison, S., quoted, 16
Alliteration, 173
Alliterative excess, 180–182
Ambiguity, 10
  in polygonal form, 21, 41, 42, 44
  in rectangles, 29, 30
  in triangles, 21
Anastomosis, 174, 175
Angle of rotation, 18
Architecture, 202, 214
Aristotle, 191, 194, 195, 197
  quoted, 193, 194
Art, and aesthetics, 3, chap. 9
  as expressive, 192, 193, 194, 197, 200, 201, 206, 207, 217
  as material of aesthetics, 3, 14
  creative, 14, 216, 217
  Eastern view of, 207, 208
  evolution in, 215
  extent of, 209
  formal side of, 13, 14, 215
  Greek view of, 194, 195
  theories of, 216
Associations, 6, 7
  by contiguity, 8
  formal and connotative, 8, 9
  intuitive nature of, 7
  rôle of, 6, 7
  tone of feeling of, 10
  verbal, 7
Assonance, 172
Assonantal excess, 180–182
Attention, rôle of, 4, 5
Augmented fourth, 99
Automatic motor adjustments, 4–6
Axis of symmetry, 17, 18

INDEX

Bach, J. S., 127
Balance, 9
Baring-Gould, S., 188
Beautiful, 192, 193
   Aristotle quoted on, 193
   Euler on, 199
   Hemsterhuis quoted on, 4, 170, 199
   Plato quoted on, 193
   Ross quoted on, 207
Beethoven, L. von, 156–160, 165
Bernoulli, J., quoted, 51
Bissell, A. D., 157
Boileau, N., 197
Bosanquet, B., quoted, 194, 195
Bourgoin, J., 50
Browne, Sir Thomas, quoted, 189
Bruce, S., 80
Burke, E., 199
Butler, S., quoted, 171

Cacophony, in music, 163
   in poetry, 181
Cadence, authentic, 135, 136
   false, 136, 137
   half, 163
   interrupted, 135
   plagal, 135, 136
Cadential sequences, 135–137
Caskey, L. D., 67
Center of area, 20
Center of interest, 9, 10, 213
Center of symmetry, 17, 18
Characteristic network of vases, 71
Characteristic points and directions of vases, 69, 70
Cherubini, M. L., 127
Chord, of the diminished seventh, 107
   of the leading note, 103, 104, 107
Chordal sequences, 128, chap. 6
   aesthetic formula for, 143–145
   aesthetic measure of, 146–151
   bass leaps in, 142, 143
   close harmony in, 138, 140
   dissonant leaps in, 130, 131, 142
   dominant, 134, 137, 138
   elements of order in, 129, 135–143
   illustrative set of, 148–151
   leading note in, 141, 142
   limitation of leaps in, 129–131
   limitation of similarity in, 129, 132, 133
   mediant chord in, 140, 141
   problem of, 128
   progression in, 138, 139
   Prout's rating of, 145–147
   regular, 129
   similar motion in, 142
   stationary notes in, 142
   transition value in, 129
   types of root progressions in, 138, 140
Chords, 91, 92, 97–108, chap. 5
   aesthetic formula for, 108–113
   aesthetic measure of, 146–151
   ambiguous, 108
   classification of, 107
   close harmony in, 138, 140
   consonant, 103, 104
   derivatives of, 106, 124–126
   diminished fifth in, 110, 111
   dominant, 104–106, 119–124
   dominating note in, 111, 112
   elements of order in, 108–112
   expected third in, 110, 111
   fundamental positions of, 101–106, 113–116
   incomplete, 108, 116
   inversions of, 101–106, 116–119
   irregular, 105–107, 126, 127
   major, 103, 104
   minor, 103, 104
   of major character, 102–106
   primary, 103, 104
   regular, 105–107
Chromatic scale, 95, 96
Circle, 16, 18, 52
Close harmony in chordal sequences, 138, 140
Cohen, G., 204
Coleridge, S. T., 185
   quoted, 184, 186, 187
Collins, W., quoted, 181
Color in painting, 213, 214
Complexity of aesthetic object, 3, 4
   measure of, 5, 6
   physiological correlative of, 4, 5
   psychological meaning of, 5, 6
Composition in painting, 11, 212, 213
Connotative factors or elements of order, 8, 9
   in music, 88
   in ornaments, 56, 64
   in poetry, 170
   in polygonal form, 44, 45
   in vase form, 70
Consecutive octaves, 132
Consecutive perfect fourths and fifths, 132
Consonance, and dissonance, 8, 89, 90, 92
   Euler's theory of, 198, 199
   Pythagorean law of, 90
Consonantal excess, 174, 175, 183
Contrast, 9
   harmonic, 157, 162
   melodic, 157, 162
Correggio, 212

INDEX

Croce, B., 195, 206, 207
    quoted, 197, 206, 217
Crousaz, J. P. de, quoted, 197
Curvature of contour of vases, 73, 74

Decorative design, 211, 212
Derivatives of chords, 106, 124–126
Descartes, R., quoted, 196
Diatonic scale, 92–95
Diminished fifth, 99, 100
Directions, polygonal, 21, 22, 26, 41, 44
Dissonance, and consonance, 8, 89, 90, 92
    resolution of, 133, 134
Dissonant leaps, 130, 131
    in chordal sequences, 142
    in melody, 155
Dominant chord, 101–104
Dominant 7th chord, 104, 105
Dominant 9th chord, 104, 105
Dominant 11th chord, 105, 106
Dominant 13th chord, 105, 106
Dominant chordal sequences, 134, 137, 138
Dominant note, 93
Dominating note in chord, 111, 112

Effort, economy of, 202
    in aesthetic experience, 3–6, 12
Elements of order, 9
    connotative, 9, 13, 14
    density of, 4
    formal, 9, 10
    in chordal sequences, 129, 135–143
    in chords, 108–112
    in decorative design, 211, 212
    independent, 15
    indices of, 10
    in harmony, 135–143
    in melody, 160–164
    in music, 87–99
    in poetry, 178–183
    in polygonal form, 10, 11, 33–44
    in simple rectilinear ornaments, 57–62
    in vase form, 68, 69, 71, 72, 76–78, 85, 86
Ellis, A. J., 199
Empathy, 6, 72, 206
Equality, 9
Equilibrium, of polygons, 35, 36
    of simple rectilinear ornaments, 57, 58
Euler, L., 198, 199
Evolution, in art, 46, 215
    in music, 88
Expected third in chord, 110
Experimental melodies, viii, 165–168
Experimental poem, viii, 185, 186
Experimental vases, viii, 84, 85

Fechner, G. T., 28, 30, 202, 205
Fracastoro, G., 195
Full tone, 94
Fuller, T., quoted, 170
Fundamental portion of ornament, 52
Fundamental positions, of chords, 103–106, 113–116
    of triads, 102, 103
Fundamental region of ornament, 52

Generators of chords, 102, 103–106
Genius, 216
Giles, H. A., 208
Goblets, 68, 71
Goethe, J. W. von, 193
Golden Rectangle, 24, 27–31
    Lipps quoted on, 29, 30
Golden Section, 27, 195
Graun, K. H., 124
Greene, R., quoted, ix
Group of motions, of ornaments, 51–55
    of polygons, 36, 37
Guest, E., quoted, 188
Gurney, E., 97, 206
    quoted, 206

Hambidge, J., 67, 72
Handel, G. F., 127
Hanslick, E., quoted, 204
Harmonic contrast, 157, 162
Harmonic sequence, 160, 162
Harmony, chaps. 5, 6
    elements of order in, 135–143
    problem of, 89, 97, 98, 128
Hegel, G. W. F., quoted, 200
Helmholtz, H. von, 8, 92, 202, 203, 204
    quoted, 199, 203
Hemsterhuis, F., quoted, 4, 170, 199
Herbart, J. F., 200, 201
Hidden octaves, 132, 133
Hidden perfect fourths and fifths, 132, 133
Hobson, R. L., 79
    quoted, 78
Hogarth, W., 199
Hsieh Ho, Six Canons of, 207, 208

Illusions, 47, 48
Indices, of automatic motor adjustments, 5
    of associations, 10
Intervals, musical, 98–101
    perfect, 97
Inversions of chords, 101–106, 116–119

Jowett, B., 22

Kant, I., 200
    quoted, 200
Keats, J., quoted, 188

INDEX

Lamb, C., quoted, 185
Lanier, S., 171, 205, 206
  quoted, 174, 205, 206
Leading note, 94
  chord of the, 107
  in chordal sequences, 141, 142
  in melody, 163
Leaps, in harmony, 129–131
  in melody, 155, 163
Leibnitz, G. W., 196, 197
  quoted, 197
Lipps, T., 6, 72, 206
  quoted, 29, 30
Longfellow, H. W., quoted, 187
Lowell, Amy, 188, 189
  quoted, 187
Lowes, J. L., 181, 216

Major character of chords, 101–106
Major mode, 96, 97
Major second, 98
Major seventh, 99
Major sixth, 99
Major third, 94, 97, 99
Mathematics, and aesthetics, 46, 47
  and music, vii, 204
Mediant chord, 102–104, 115–119, 140, 141
Mediant note, 94
Melodic contrast, 157, 162
Melodic sequence, 157, 162
Melody, 88, 89, chap. 7
  aesthetic formula for, 160–164
  aesthetic measure of, 156–160, 164–168
  cadence in, 156, 158, 161
  comparison in, 154, 155, 161, 162
  complexity of, 152, 153, 162
  continuity in, 163
  elements of order in, 160–162
  embellishment of, 156, 164
  freedom from formal blemishes in, 163
  harmonic contrast in, 157, 162
  harmonic sequence in, 160, 162
  harmony and, 153
  inversion in, 157, 162
  leading note in, 163
  limitation of leaps in, 155, 163
  melodic contrast in, 157, 162
  melodic sequence in, 157, 162
  phrasing in, 154, 155
  problem of, vii, 152, 153, 206
  regularity of pattern in, 163
  repetition in, 154, 157, 158, 161, 163
  rhythm and, 169
  riddle of, vii
  satisfactory form in, 163, 164
  secondary, 153, 159, 160, 162
  tonic close in, 159, 160
  tonic start in, 156, 157, 160
  transposition in, 158, 161
Mendelssohn, J. L, F., 129
Metre, 170, 171
Michelangelo, 67, 195
Minor mode, 96, 97
Minor second, 98
Minor seventh, 99
Minor sixth, 99
Minor third, 97, 99
Modulation, 96
Morrey, C. B., 164
Music, Eastern, 214, 215
  emotional effect of, 87, 88, 202
  evolution of, 88
  structure of, 152
  Western, 87, 214
Musical form, Hanslick quoted on, 204
  problem of, vii, 87–89
Musical notes, 7, 8, 89–92
Musical vowels, 173, 174

Networks, 11, 32, 33, 39–41, 46
Niches, polygonal, 32, 41, 44
Normal observer, 11

Onomatopoeia, rôle of, 170, 172
Optical illusions, 47, 48
Order of aesthetic object, 3, 4
  psychological meaning of, 10, 11
Ornamental patterns, 55, 56, 64, 211, 212
Ornaments, 49–51, chap. 3
  aesthetic formula for, 57–62
  aesthetic measure of, 62, 63
  aesthetic problem of, 56
  and ornamental patterns, 55, 56, 64
  classification of, 52–55
  curvilinear, 52, 55, 63–64
  elements of order in simple rectilinear, 51, 57–62
  fundamental portions of, 51–54
  fundamental regions of, 52–54
  group of motions of, 51, 52
  mixtilinear, 52
  motions of, 49
  one-dimensional, 52–55
  polygons as, 57
  rectilinear, 52, 55, 62, 63
  simple, 52, 62, 63
  species of, 54, 55
  two-dimensional, 53–55

Paciolo, L., 27, 195
Painting, 11, 212–214
  centers of interest in, 213

INDEX

color in, 213, 214
composition in, 11, 212, 213
elements of order in, 207
Six Canons of Hsieh Ho in, 207, 208
Phonetic analysis, 177, 178
Phonetic syzygy, 176, 177
Piston, W. H., 160, 165–168
Pitch, 89, 90
Plato, viii, 22, 28, 191–195, 197
    quoted, 22, 193
Plotinus, quoted, 194
Poe, E. A., 177, 178, 182, 185, 201, 202, 203, 206
    his concept of poetry, 175, 176
    quoted, 172, 173, 175, 180, 188, 201, 202
Poetry, chap. 8
    aesthetic formula for, 183, 184
    aesthetic measure of, 184–189
    alliteration in, 173, 178–180
    alliterative excess in, 180–182
    anastomosis in, 174–176
    assonance in, 172, 173, 178–180
    assonantal excess in, 180–182
    cacophony in, 181, 184
    complexity of, 174, 175, 177, 178
    consonantal excess in, 174–175, 183
    elements of order in, 178–183
    in other languages, 189
    licenses in, 170
    metre in, 170–171
    musical quality in, 170–175, 189, 190
    musical vowels in, 173, 174, 182
    onomatopoeia in, 170
    phonetic analysis of, 177, 178
    phonetic syzygy in, 176, 177
    Poe's concept of, 174, 175
    rhyme in, 172, 178
    significance in, 170
    Sylvester's concept of, 176, 177
Polyá, G., 55
Polygonal forms, 10, 11, 16, chap. 2
    aesthetic formula for, 42–44
    aesthetic measure of, 44–46
    aesthetic problem of, 16, 17
    Alison quoted on, 16
    ambiguity in, 21, 41, 42, 44
    and networks, 11, 32, 33, 39–41, 46
    complexity of, 6, 32, 34
    connotations in, 44, 45
    decorative use of, 16
    directions of, 21, 22, 26, 41, 44
    elements of order in, 10, 11, 33–44
    equilibrium of, 35, 36
    group of motions of, 36, 37
    niches in, 32, 41, 44
    Plato quoted on, 22

    square as most beautiful of, 46, 47
    symmetry in, 17, 18
Pope, Alexander, quoted, 170
Pope, Arthur, quoted, 207, 210–213
Pottier, E., quoted, 67
Primary notes, 93, 94
Progression in chordal sequences, 138, 139
Prose, sonorous, 189
Prout, E., 132, 133, 155
    his rating of chordal sequences, 145–147
    quoted, 114 *et seq.*, 140, 146
Puzzle-art, 216
Pythagoras, 90

Quadrilaterals, 23–27
Quasi-ornaments, 50, 51

Rameau, J. P., 198, 203
Rectangles, 17, 18
    comparison of, 27–31
    Golden, 27–30
    in composition, 30, 31
    Lipps, quoted on, 29, 30
    special, 28
Regular chords, 105–107
Repetition, 9
    in architecture, 214
    in melody, 154, 157, 158, 161, 163
    undue, 10
Resolution of dissonance, 133, 134
Rhyme, 172, 178
Rhythm, in Eastern and Western music, 214, 215
    the problem of, 152, 169
Roots, of chords, 103–105
    of triads, 102, 103
Ross, D. W., 207, 214
    quoted, 207, 211

Scale, 92, 93
    chromatic, 95, 96
    equally tempered diatonic, 94, 95
    natural diatonic, 92–94
    pentatonic, 214
    quarter-tone, 214
Schiller, F., quoted, 200
Schleiermacher, F., quoted, 201
Sculpture, 214
Secondary melody, 153, 159, 160, 162
Semitone, 94
Sensuous feeling, 7, 8
Sequence, 9
Sequence of first inversions, 138
Shakespeare, W., quoted, 187
Shelley, P. B., quoted, 170, 171
Similarity, 9

INDEX

and spiral forms, 50, 51
    in ornaments, 51, 57–59
Similar motion in chordal sequences, 142
Socrates, 192
Species of ornaments, 54, 55
Speiser, A., 53, 55, 70
Spencer, H., 202
Spirals, 50, 51
Square, 16–18, 23–25
    as most beautiful polygon, 46, 47
Stationary notes in chordal sequences, 142
Style, 202
Subdominant chord, 101–104
Subdominant note, 93
Submediant chord, 107
Submediant note, 94
Supertonic chord, 107
Supertonic note, 94
Sylvester, J. J., 175, 177, 203, 204
    his concept of poetry, 176, 177
    quoted, 173, 174, 176, 203, 204
Symmetry, 9, 23
    appreciation of, 7
    axis of, 17, 18
    center of, 17, 18
    central, 18
    in Japanese art, 23
    in polygonal form, 17, 18, 35, 36–39
    rotational, 17, 18, 36–39

Taj Mahal, 214
Takeuchi Seiho, 23
Taste, 195, 196
Tennyson, A., 179
    quoted, 171, 180, 187
Tilings, 50, 64
    aesthetic, formula for, 64–65
    aesthetic measure of, 64–66
    fundamental visual field of, 64–65
    interesting polygons in, 64–65
Tonality, 95, 96, 99–101
Tone of feeling, 5, 7, 8, 10
    indices of, 10
Tonic chord, 101–104
Tonic close, 159, 160
Tonic note, 93, 95, 96

Tonic start, 156, 157, 160
Transposition in melody, 158, 161
Triads, 101–103
Triangles, 16, 19–23
    comparison of, 21, 22
    equilateral, 16, 18, 19, 20
    in Japanese art, 23
    isosceles, 16, 18, 19, 20
    Plato on, 22
    scalene, 16, 21, 23
Tweedy, D. N., 139

Uncertainty, 47, 48
Unison, 98
Unity in variety, 4, 194, 195
Unnecessary imperfection, 10, 25

Valéry, P., quoted, 190
Vases, chap. 4
    aesthetic formula for, 78
    aesthetic measure of 8 Chinese, 79–84
    aesthetic measure of 3 experimental, 84–85
    aesthetic problem of, 67–68
    centers of, 72
    characteristic network of, 71
    characteristic points and tangents of, 69–71
    Chinese, 78
    complexity of, 76
    conventional requirements for, 74, 75
    curvature of contour of, 73, 74
    elements of order in, 68, 69, 71, 72, 76–78, 85, 86
    empathetic theory of, 72
    experimental, 84, 85
    geometric configuration of, 71
    Greek, 67, 78
    interpretation of form of, 76
    regularity of contour of, 73, 74
    symbols of, 69, 76
    utilitarian requirements for, 74, 75
Veronese, 212, 213
Vico, G. B., 197, 198
Voltaire, F. M. A. de, quoted, 170

Wit, 195, 196